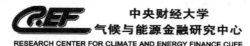

2018 中国气候融资报告
2018 China Climate Financing Report

崔 莹　洪睿晨　著

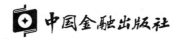

责任编辑：肖　炜　董梦雅
责任校对：张志文
责任印制：张也男

图书在版编目（CIP）数据

2018中国气候融资报告＝2018 China Climate Financing Report / 崔莹，洪睿晨著．—北京：中国金融出版社，2019.8
ISBN 978 - 7 - 5220 - 0121 - 0

Ⅰ．①2018…　Ⅱ．①崔…　②洪…　Ⅲ．①气候变化—融资—研究报告—中国—2018　Ⅳ．①P467 - 05

中国版本图书馆CIP数据核字（2019）第102574号

2018中国气候融资报告
2018 Zhongguo Qihou Rongzi Baogao

出版
发行　中国金融出版社
社址　北京市丰台区益泽路2号
市场开发部　（010）63266347，63805472，63439533（传真）
网上书店　http：//www.chinafph.com
　　　　　（010）63286832，63365686（传真）
读者服务部　（010）66070833，62568380
邮编　100071
经销　新华书店
印刷　北京市松源印刷有限公司
尺寸　169毫米×239毫米
印张　13.5
字数　214千
版次　2019年8月第1版
印次　2019年8月第1次印刷
定价　50.00元
ISBN 978 - 7 - 5220 - 0121 - 0
如出现印装错误本社负责调换　联系电话（010）63263947

关于本报告

中央财经大学绿色金融国际研究院（IIGF）

中央财经大学绿色金融国际研究院（以下简称绿金院）是国内首家以推动绿色金融发展为目标的开放型、国际化的研究院。绿金院前身为中央财经大学气候与能源金融研究中心，是中国金融学会绿色金融专业委员会的常务理事单位。绿金院以营造富有绿色金融精神的经济环境和社会氛围为己任，致力于打造国内一流、世界领先的具有中国特色的专业化金融智库。

中央财经大学气候与能源金融研究中心（RCCEF）

中央财经大学气候与能源金融研究中心成立于2011年9月，已连续八年发布《中国气候融资报告》，基于广义的全球气候融资概念，形成了气候融资流的分析框架，构建了中国气候融资需求模型，并从国际气候资金治理以及中国气候融资的发展角度，逐年进行深入分析，积累了一系列的气候融资研究成果。基于长期信任与合作的基础，中央财经大学气候与能源金融研究中心与财政部建立了部委共建学术伙伴关系。

指导

王 遥	中央财经大学绿色金融国际研究院院长、教授、博导
	中国金融学会绿色金融专业委员会副秘书长

作者

崔 莹	中央财经大学气候与能源金融研究中心高级研究员
洪睿晨	中央财经大学气候与能源金融研究中心研究员

2018中国气候融资报告
2018 China Climate Financing Report

Mathias Lund Larsen	中央财经大学气候与能源金融研究中心研究员
崔连标	中央财经大学绿色金融国际研究院特邀研究员
张诗雨	中央财经大学气候与能源金融研究中心研究助理
田晓晔	中央财经大学气候与能源金融研究中心研究助理
钱青静	中央财经大学气候与能源金融研究中心研究助理

前　言

2018年是继2015年《巴黎协定》通过后全球应对气候变化行动取得重大进展的一年。联合国政府间气候变化委员会（IPCC）发布《IPCC全球升温1.5℃特别报告》（以下简称《1.5℃报告》），评估了全球气温上升1.5℃和2℃的气候影响，并提出了可能的减排路径。《1.5℃报告》向大众描绘了将升温控制在1.5℃内的更美好的世界，然而联合国环境规划署（UNEP）公布的《排放差距报告2018》（*Emissions Gap Report* 2018）却向世人发出警告，2017年全球温室气体排放再创492亿吨二氧化碳当量新高。从减排技术角度分析，达成1.5℃温控目标的概率正在减小，而要达成2℃温控的目标，各国还需付出比现在的减排工作更为积极的努力。

2018年出现了一些对应对气候变化行动不利的事件，如法国巴黎爆发了抗议政府加征燃油税的"黄背心"运动，瑞士议会投票反对以达到《巴黎协定》目标为目的的二氧化碳法案修正案，德国煤炭退出委员会建议将关闭第一批燃煤电厂的期限从2020年推迟至2022年，越来越多的证据显示，部分欧盟国家很大概率上无法完成2030年的减排目标。

在这些负面影响下召开的第24次卡托维兹联合国气候变化大会可谓是让全球应对气候变化行动重整旗鼓。与会各方最终对《巴黎协定》大部分内容的实施细则达成一致，重申了发达国家对气候资金的出资义务。

中国对此次卡托维兹联合国气候变化大会谈判结果的促成起到了很大的推动作用。中国代表团在各方代表之间调解斡旋，有效地推动了谈判进程，并贡献了气候治理方面的中国经验。在美国退出《巴黎协定》的背景下，中国在国际气候治理中越来越发挥"引领者"的作用，承担起了大国责任。

中国近年来开展的"一带一路"建设倡导低碳发展、绿色投资。"一带一路"沿线国家生态环境脆弱、对气候变化应对能力较差，进行气候投融资不仅可以满足当地人民的生存需求，提高生活质量，还能用绿色产能有效淘汰高能耗高污染产能，防止环境进一步恶化，降低气候风险。

与此同时，中国国内气候融资也在稳步发展。2017年底中国宣布启动全国碳市场建设，目前正处于基础建设期。中国碳市场一旦建成将成为全球最大体量的碳市场，减排效应显著。而以绿色债券、绿色信贷为代表的绿色金融体系的蓬勃发展也会为中国应对气候变化提供气候资金和示范效应。

总体来看，距离《巴黎协定》约定的2020年后全球应对气候变化行动的时间截点已经很近，然而国际社会对各方应对气候变化的任务分工依然不明朗，各国政府还需继续努力，持续推动国际谈判进展和加强本国行动。鉴于中国在绿色金融领域的领先经验和国际影响力的提升，未来必将成为全球气候治理中举足轻重的角色，影响气候融资的整体发展。

英文缩写索引表

英文简写	英文全称	中文对应名称
A		
ADB	Asian Development Bank	亚洲开发银行
AfDB	African Development Bank	非洲开发银行
AIIB	Asian Infrastructure Investment Bank	亚洲基础设施投资银行
B		
BOC	Bank of China	中国银行
BFIs	Bilateral Financial Institutions	双边金融机构
C		
CCER	China Certified Emission Reduction	中国核证自愿减排量
CDB	China Development Bank	国家开发银行
CDM	Clean Development Mechanism	清洁发展机制
CDMF	Clean Development Mechanism Fund	清洁发展机制基金
CER	Certification Emission Reduction	核证减排量
CIFs	Climate Investment Funds	气候投资基金
COP	Conferences of the Parties	联合国气候变化框架公约缔约方大会
CPI	Climate Policy Initiative	气候政策倡议
D		
DFIs	Developmental Finance Institutions	发展性金融机构
E		
EBRD	European Bank for Reconstruction and Development	欧洲复兴开发银行
EC	European Commission	欧盟委员会
EIB	European Investment Bank	欧洲投资银行
ERPA	Emission Reduction Purchase Agreement	减排量购买协议
EU－ETS	European Union Emission Trading System	欧盟碳排放交易体系

续表

英文简写	英文全称	中文对应名称
EXIMB	the Export – Import Bank of China	中国进出口银行
F		
FAO	Food and Agriculture Organization	联合国粮食和农业组织
G		
GCA	Global Commission on Adaptation	全球适应委员会
GCF	Green Climate Fund	绿色气候基金
GDP	Gross Domestic Product	国内生产总值
GEF	Global Environmental Facility	全球环境基金
GGGI	Global Green Growth Institute	全球绿色发展署
I		
IADB	Inter – American Development Bank	泛美开发银行
IBRD	International Bank for Reconstruction and Development	国际复兴开发银行
ICBC	Industrial and Commercial Bank of China	中国工商银行
IEA	International Energy Agency	国际能源署
IFC	International Finance Corporation	国际金融公司
INDC	Intended Nationally Determined Contributions	国家自主贡献目标
IPCC	Intergovernmental Panel on Climate Change	联合国政府间气候变化委员会
IRENA	International Renewable Energy Agency	国际可再生能源机构
J		
JI	Joint Implementation	联合履约机制
L		
LDCs	Least Developed Countries	最不发达国家
M		
MDBs	Multilateral Development Banks	多边开发银行
MFIs	Multilateral Financial Institutions	多边金融机构
MIGA	Multinational Investment Guarantee Agency	多边投资担保机构
MRV	Monitoring, Reporting and Verification	国际气候资金核查、报告和监测
N		
NDBs	New Development Bank	新开发银行

续表

英文简写	英文全称	中文对应名称
O		
ODA	Official Development Assistance	官方发展援助组织
OECD	Organization for Economic Co-operation and Development	经济合作与发展组织
P		
PSF	Private Sector Facility	私营部门机制
PPP	Public – Private Partnership	政府与社会资本合作模式
R		
REDD +	Reducing Emissions from Deforestation and Degradation	减少毁林和森林退化造成的温室气体排放
S		
SPV	Special Purpose Vehicle	特殊目的实体
U		
UNDP	United Nations Development Programme	联合国开发计划署
UNEP	United Nations Environment Programme	联合国环境规划署
UNFCCC	United Nations Framework Convention on Climate Change	联合国气候变化框架公约
UN-REDD	United Nations Reducing Emissions from Deforestation and Forest Degradation	减少发展中国家毁林和森林退化所致排放量联合国合作方案
W		
WB	World Bank	世界银行
WBG	World Bank Group	世界银行集团
WMO	World Meteorological Organization	世界气象组织
WRI	the World Resource Institution	世界资源研究所

目 录
Contents

一、全球气候融资进展 ··· 1
 （一）全球气候资金量有所上升，但仍存在巨大资金缺口 ········ 2
 （二）发达国家对发展中国家的气候资金承诺尚待落实 ········· 5
 （三）可再生能源气候融资呈增长趋势，但仍不能满足温控要求 ······ 7
 （四）区块链技术开始在气候融资领域中应用 ·················· 10

二、中国气候融资进展 ··· 12
 （一）全国碳市场宣布启动，地方碳试点持续运行 ············· 12
 （二）绿色金融对气候融资的推动力不断显现 ·················· 18
 （三）PPP模式在生态环保领域稳步推进 ······················ 23
 （四）中国积极参加国际交流合作，为全球应对气候变化作贡献 ······ 29

三、"一带一路"沿线的气候投融资 ································ 34
 （一）"一带一路"倡议的基本情况 ···························· 35
 （二）"一带一路"沿线主要国家的气候特征 ···················· 36
 （三）推动"一带一路"沿线绿色投资的基础 ···················· 40
 （四）中国对"一带一路"沿线国家的气候投融资项目 ··········· 43
 （五）进行"一带一路"气候投融资的主要机构 ················· 46

四、绿色气候基金（GCF）的发展情况和融资方案分析 ·············· 54
　（一）GCF 的发展情况 ·· 54
　（二）GCF 融资责任的分摊机制分析 ······································· 58

五、多边开发银行在气候融资中的角色 ······································· 65
　（一）多边开发银行在促进可持续发展方面具有优势 ················· 65
　（二）气候融资发展面临多种挑战 ··· 67
　（三）多边开发银行在气候融资中起重要作用 ··························· 68
　（四）应采用多种方式继续推进气候融资 ································· 70

六、政策建议 ·· 73
　（一）在全球气候治理格局发生改变的新形势下继续发挥"引领者"
　　　　作用 ·· 73
　（二）积极利用国际气候资金推动国内减缓和适应气候变化 ········ 74
　（三）大力推动市场化碳交易并发展碳金融 ····························· 76
　（四）加强"一带一路"沿线的气候投融资 ····························· 77

一、全球气候融资进展

2015年12月《巴黎协定》通过后,全球一直未就《巴黎协定》的实施细则达成一致。2018年12月,第24届联合国气候变化缔约方大会于波兰卡托维兹召开,与会各国经过两周的谈判,最终就应对气候变化的行动细则达成一致,制定了《巴黎协定》大部分内容的实施细则,为进一步实现《巴黎协定》目标奠定了基础。大会重申了在2020年以后发达国家向发展中国家每年至少动员1000亿美元的资金支持。

在大会召开前,联合国政府间气候变化委员会(IPCC)发布的《IPCC全球升温1.5℃特别报告》,指出目前全球气温较工业化前水平已经增加了1℃,全球升温1.5℃最快有可能在2030年达到。全球温升1.5℃与2℃的气候影响差异显著,当温升到达2℃时将带来更具破坏性的后果,如栖息地丧失、冰盖融化、海平面上升等,威胁人类的生存和发展,也将给世界经济带来更大损害。为实现1.5℃温控目标,全球气候行动亟待加速。联合国环境规划署(UNEP)也发布了《排放差距报告2018》(*Emission Gap Report 2018*),指出目前形势下,距离《巴黎协定》设定的全球升温不超过2℃目标仍有较大差距。若按当今的发展态势,《巴黎协定》温控目标必然失败,且即便各国都按理想状况履行自己的国家自主贡献目标(INDC)中的减排计划,地球均温仍可能在21世纪末上升超过3℃。两份报告都表明,气候谈判必须加大力度,促使各国作出更严格的减排行动。

气候融资是促进全球减排的重要途径,目前全球气候资金流有所上升,但资金缺口仍然巨大;发达国家对发展中国家的"1000亿美元"气候资金承诺尚待落实;可再生能源气候融资呈增长趋势,但仍不能满足温控要求;另外,新兴的区块链技术开始在气候融资领域应用。

（一）全球气候资金量有所上升，但仍存在巨大资金缺口

2018年11月底，联合国气候变化框架公约（UNFCCC）在波兰卡托维兹气候变化大会召开前发布了一份全球气候资金流双年报（2018 *Biennial Assessment and Overview of Climate Finance Flow*），提供了2015年和2016年通过各途径可统计的气候资金流的情况。

根据UNFCCC统计，气候资金数据的质量和完整性从2016年以来有所提高，各机构改进了评估气候资金数据的方法。全球气候资金流动整体增量显著，在可比较的基础上，2015—2016年全球气候资金流数额较2013—2014年增加了17%，从2014年的5840亿美元上升到2015年的6800亿美元，再到2016年的6810亿美元。2015年的快速增长很大程度上是由可再生能源方面新投入的私人资金推动的，占据了总金额中最大的部分。2016年可再生能源方面投入的资金降低，原因是可再生能源技术成本的持续降低和新项目需要的资金融资，但这方面的减少被建筑、工业和交通领域能源效率技术新增的8%的资金投入所抵消，使2016年的气候资金流较2015年略有增长。

1. 从附件Ⅱ国家流向非附件Ⅰ国家的气候资金情况[①]

附件Ⅱ国家提交的关于气候融资的专项报告中显示，气候资金投资数量和增长率均较前两年有所增加。2013年至2014年气候资金总额仅增加了5%，但从2014年至2015年增加了24%，达到330亿美元。随后又增加了14%，2016年总额达380亿美元。在这些资金中，2015年有300亿美元和2016年有340亿美元是通过双边、区域和其他渠道提供的；剩下的是通过多边渠道提供的。从2014年至2016年，减缓和适应气候资金的增长比例大致相等，分别为41%和45%（见图1-1）。

[①] 《联合国气候变化框架公约》将缔约方分为三类，分别为附件Ⅰ缔约方、附件Ⅱ缔约方以及发展中国家缔约方。附件Ⅰ缔约方由24个OECD国家、欧洲共同体成员以及11个向市场经济过渡的国家组成，附件Ⅱ缔约方由24个OECD国家及欧洲共同体成员组成。具体来说，附件Ⅱ国家包括澳大利亚、意大利、奥地利、日本、比利时、卢森堡、加拿大、荷兰、丹麦、新西兰、欧洲共同体、挪威、芬兰、葡萄牙、法国、西班牙、德国、瑞士、希腊、土耳其、瑞典、冰岛、大不列颠北爱尔兰联合王国、爱尔兰和美利坚合众国。附件Ⅰ国家除了这些外，还包括爱沙尼亚、白俄罗斯、保加利亚、捷克、斯洛伐克、匈牙利、罗马尼亚、俄罗斯、拉脱维亚、立陶宛、乌克兰和波兰。

一、全球气候融资进展

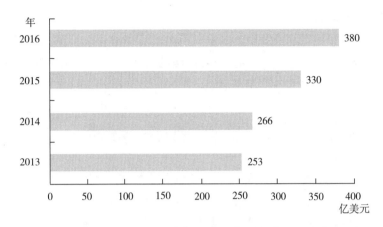

图1-1　2013—2016年从附件Ⅱ国家流向非附件Ⅰ国家的气候资金

2. 多边气候资金流向情况

2015年和2016年通过UNFCCC基金和多边气候基金提供的资金总额分别为14亿美元和24亿美元，资金总额的大幅度增长是绿色气候基金加强运作的结果。总体而言，这与2013年至2014年相比减少了约13%，是因为气候投资基金（CIFs）减少了承诺的资金投入，以及绿色气候基金（GCF）在2016年才开始扩大运营规模（见图1-2）。

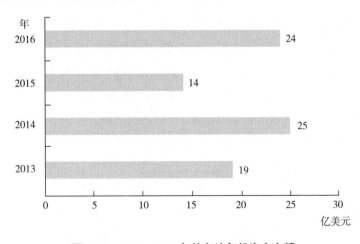

图1-2　2013—2016年的多边气候资金金额

3. 多边开发银行提供的气候资金

多边开发银行分别在2015年和2016年从其自身资源向符合条件的受援国提供了234亿美元和255亿美元的气候资金。这比2013—2014年平均增加

3

了3.4%（见图1-3）。

多边开发银行气候资金中，由除韩国外的OECD-DAC成员提供给符合官方发展援助组织（ODA）要求的受援国的气候资金总额最多为2015年的174亿美元和2016年的197亿美元。

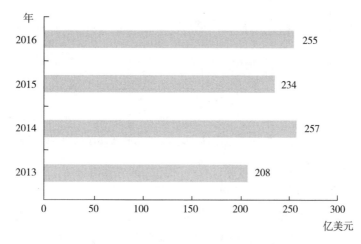

图1-3 2013—2016年多边开发银行从自身资源提供的气候资金金额

4. 私营部门气候资金

私营部门数据的地理归属存在很大的不确定性。虽然多边开发银行和经济合作与发展组织自2016年以来一直致力于通过多边和双边机构调动私营气候资金，但关于私营部门气候资金的数据来源和投向地仍不清晰。

多边开发银行报告显示，2015年多边开发银行调动的私营部门资金为109亿美元，2016年增加了43%，达到157亿美元。据OECD统计，2012年至2015年，双边和多边机构调动的私营部门气候资金为217亿美元，其中多边渠道为140亿美元，双边渠道为77亿美元。

5. 国家内部气候资金流动情况

国家和地方政府的国内气候融资是全球气候资金流的一个潜在来源，特别是在某些情况下，国家自主贡献被转化为具体的投资计划，且国内监测和跟踪气候支出的力度也在加强。2015—2016年国内公共气候融资估计数达到670亿美元。然而，关于国内气候支出的全面数据并不容易获得，因为这些数据不是定期收集的，而且国家内部或国家之间采用的收集方法也不统一。

6. 非OECD-DAC国家、有资格获得官方发展援助的国家和非附件Ⅰ国家间气候资金流向

非附件Ⅰ国家间气候资金流动的信息没有得到系统的跟踪，主要靠这些国家通过OECD-DAC报告系统自愿上报，以及通过非OECD国家的发展金融机构自主披露。2015年此类资金流动总额为122亿—139亿美元，2016年为113亿—137亿美元，比2013—2014年平均增长约33%。新的多边机构包括亚洲基础设施投资银行（AIIB）和新开发银行（NDBs），在2016年共同向可再生能源项目提供了9.11亿美元。虽然气候资金流总体上在2015—2016年双年期比上一个双年期有所增加，但要满足减缓和适应气候变化所需的最低投资额，这一资金水平还远远不够。国际能源署（IEA）的预测表明，为实现《巴黎协定》的温控目标，能源领域在2015—2030年需投入资金16.5万亿美元。全球绿色发展署（GGGI）估计2016—2030年的气候融资缺口为2.5万亿—4.8万亿美元。根据联合国政府间气候变化专家小组（IPCC）的统计，2010年—2050年发展中国家适应资金需求量为700亿—1000亿美元每年。据UNEP估计，到2030年适应资金的年需求量为1400亿—3000亿美元每年；到2050年，适应资金的年需求量将增至2800亿—5000亿美元。尽管当前对私人和公共资源所需气候投资的估计存在较大差异，但各机构的统计数据均传递了相同信息：全球气候资金所需的投资金额远远大于现今已投入的数量。

（二）发达国家对发展中国家的气候资金承诺尚待落实

2009年在哥本哈根联合国气候变化大会上，发达国家出于历史排放原因，承诺到2020年以后每年为发展中国家提供1000亿美元的资金支持其应对气候变化行动。但由于一直未有针对这1000亿美元的具体出资定论，资金的落实情况并不理想。发达国家与发展中国家就资金是"提供"还是"动员"，是"公共的"还是"公共和私人的"，是否应为传统ODA之外"新的、额外的"等问题产生了较大分歧。

随后各年度召开的联合国气候变化大会上，1000亿美元的气候资金支持一直是气候谈判的重点。2015年12月，UNFCCC的近200个缔约方在巴黎气候大会上通过新的全球气候协定——《巴黎协定》。根据协定，在资金支持方面，在2020年以后发达国家向发展中国家每年至少动员1000亿美元的资金支持，2025年前将

确定新的数额,并持续增加。《巴黎协定》于 2016 年 11 月 4 日正式生效,1000 亿美元成为发达国家向发展中国家提供资金支持的下限。

2018 年 12 月,在波兰卡托维兹举办的第 24 届联合国气候变化大会围绕《巴黎协定》的实施细则展开谈判与磋商,为 2020 年后《巴黎协定》的实施奠定基础。卡托维兹会议为《巴黎协定》中的大部分条款落实了实施细则,在气候资金方面,重申了发达国家对发展中国家以 1000 亿美元为下限的资金支持。

整体来说,气候资金流动趋势表明,气候资金向受益国的流量在不断增加。国际公共气候资金通过三个渠道流向非附件Ⅰ国家,即多边气候基金,例如气候投资基金、绿色气候基金等;双边机构,例如法国开发署、德国复兴信贷银行等;多边开发银行,例如世界银行、欧洲投资银行、美洲开发银行等。2015—2016 年国际公共气候资金流特征见表 1-1。

表 1-1   2015—2016 年国际公共气候资金流特征

	年平均融资(亿美元)	支持的领域				融资工具		
		适应(%)	减缓(%)	REDD+①(%)	跨领域(%)	赠款(%)	优惠贷款(%)	其他(%)
多边气候基金	19	25	53	5	17	51	44	4
双边机构	317	29	50	—	21	47	52	<1
多边开发银行	224	21	79	—	—	9	74	17

数据来源:UNFCCC. 2018 Biennial Assessment and Overview of Climate Finance Flow [R]. 2018.

从表 1-1 对国际公共气候资金的总结可以看出,对减缓领域的资金支持仍然大于对其他领域的支持。对于适应领域,双边机构用于适应的比例最大,为 29%,多边气候基金和多边开发银行用于适应的比例分别为 25% 和 21%。赠款和优惠贷款是提供气候基金的主要工具。在 2015—2016 年,赠款和优惠贷款分别占了双边机构和多边气候基金总额的 95% 和 99%,通过多边开发银行的 74% 的资金是以优惠贷款的形式发放的,另外 9% 的比例是赠款。

对于发达国家履约情况的统计口径目前并无统一定论,UNFCCC 双年报

---

① REDD+ 指 "Reducing greenhouse gas Emissions from Deforestation and forest Degradation in developing countries"。指在发展中国家通过减少砍伐森林和减缓森林退化而降低温室气体排放,"+"的含义是增加碳汇。

也只是列出了各种途径的气候资金流数据,没有给出如何跟"1000亿美元"的资金承诺相对应的统计。但是从UNFCCC双年报的数据统计信息可以看出,发达国家对发展中国家的公共资金支持距离达成"1000亿美元"的出资目标仍有较大差距,即使计入私营部门资金,也仍不能达到1000亿美元。

(三) 可再生能源气候融资呈增长趋势,但仍不能满足温控要求

全球经济正在从传统的以化石能源为基础的增长模式向以可持续利用资源和能源为基础的模式转化,因此近年来关于可再生能源研究的深度和广度日益加大。根据IPCC的测算,若要将全球升温控制在1.5℃内,可再生能源发电在2050年要占全部发电量的70%—85%,煤炭发电的比例需要接近于零,天然气发电也需要使用二氧化碳的回收和储存技术。

2018年6月,21世纪可再生能源政策网络(REN21)发布了《2018可再生能源全球现状报告》,总结了2017年可再生能源的全球发展情况。该报告显示,2017年可再生能源装机增长占全球电力装机净增量的70%,比2016年提升了7个百分点。而占全球终端能源需求量80%的供热、制冷和交通领域对于可再生能源的使用却远落后于电力行业,转型动力不足。综合全球态势,可再生能源投资主要有以下四个趋势。

1. 发展中国家在可再生能源投资上占主导地位

可再生能源的新增投资中有很大一部分来自发展中国家。自2015年以来,发展中国家和新兴经济体对可再生能源的投资一直处于领先地位。中国的可再生能源投资规模巨大,2017年同比增长了30.7%。在中国的带领下,发展中国家2017年的投资额高达1770亿美元,占全球可再生能源投资的63%。而发达国家的投资额为1030亿美元,较上年度下降了19%。若以单位国内生产总值(GDP)为标准,马绍尔群岛、卢旺达、所罗门群岛、几内亚比绍和一些其他发展中国家,对可再生能源的投资远高于发达国家和新兴经济体(见图1-4)。

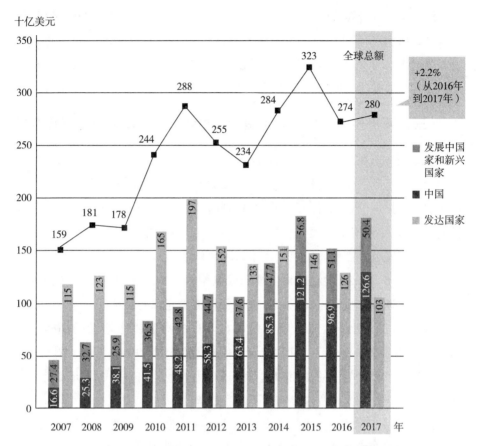

数据来源：21世纪可再生能源政策网络（REN21）. 2018可再生能源全球现状报告［R］. 2018.

图1-4 2007—2017年全球发达国家、新兴国家、发展中国家对可再生能源和燃料的新增投资

2. 光伏发电和风电占新增装机的主要份额

2017年全球可再生能源新增装机约178 GW，其中光伏发电新增约98 GW，比2016年增长了33%，占2017年新增总额的55%。新增光伏装机量大于化石燃料和核电净增装机量之和，为史上最高纪录。此外，全球风电新增装机量达52 GW，占新增总额的29%。光伏和风电的新增装机占可再生能源新增额的84%。

3. 可再生能源转型速度不如预期

据统计，GDP每增长3.7%，能源需求就会增长2.1%[①]。尽管全球对能

---

① 可再生能源政策网络（REN21）. 2018可再生能源全球现状报告［R］. 2018.

8

源转型日益重视，各国也在逐渐加大能源转型投资，但电力行业可再生能源转型速度却远不如预期。而占全球终端能源需求量80%的供暖、制冷以及交通领域，可再生能源在其中的占比还远落后于电力行业。2017年，可再生能源在全球供热能源消费中的占比仅为10.3%。而在交通领域，电力只为1.3%的交通提供了能源需求，其中又仅有约1/4是可再生能源电力。除此之外，生物燃料提供了2.8%的交通能源需求，而92%的绝大部分能源需求量仍由石油满足。目前的能源转型速度远不如预期。

4. 可再生能源政策尚需调整和改进

各国政府对可再生能源的政策不断改进以适应新的经济形势。2017年，瑞士、丹麦和越南三个国家的政府设立了新的可再生能源目标。然而，目前可再生能源政策覆盖的范围还远远不够。全球146个国家在电力行业制订了可再生能源目标，但只有48个国家制订了供暖和制冷行业的可再生能源目标，42个国家制订了交通运输领域的可再生能源目标。如果要实现《巴黎协定》中全球升温控制在工业化前水平2℃以内，并向1.5℃的目标努力，制冷、供热以及交通运输领域需要依照电力行业转型的模式，并按照更快的速度进行能源转型。

要实现能源转型，履行气候和可持续发展承诺，各国政府需要在其中发挥政策的引导作用，例如建立正确的政策框架，加快供热、制冷和交通行业的转型，推动落后领域的创新和可再生能源新技术的发展，减少甚至停止对化石燃料的补贴。

伴随全球经济增长和人口增加，能源需求持续增加。2017年能源需求和能源相关的二氧化碳排放量分别增长了2.1%和1.4%[1]。为解决能源需求的增加、能源转型等问题，国际社会对能源改革的呼声越发强烈。国际能源署（IEA）预计到2050年每年需要1万亿美元来资助能源转型。虽然气候资金流动在近十年愈加频繁，且总量呈上升态势，但仍比实际需要的低约60%[2]。国际可再生能源机构（IRENA）估计，为了减少温室气体排放并将全球平均

---

[1] 国际能源署（IEA）. 2017年全球能源和二氧化碳现状报告［R/OL］. 2017. https://www.iea.org/publications/freepublications/publication/GECO2017.pdf.

[2] Barbara K. Buchner, Padraig Oliver, Xueying Wang, Cameron Carswell, Chavi Meattle, and Federico Mazza［R］. Global Landscape of Climate Finance 2017, 2017.

温度上升限制在2℃以下，必须探索新的低碳投资模式，并在2030年前实现效益的双倍增长。低碳解决方案的部署需要将目前可再生能源的投资翻一番，到2020年达到每年5000亿美元，到2030年达到每年9000亿美元。

（四）区块链技术开始在气候融资领域中应用

区块链是指使用分散的安全数据库来支持交易中的关键步骤，即身份验证、签约和支付。参与者的身份存储在区块链上，以便在每次交易时对组织的身份进行算法验证，最大限度地降低欺诈行为的风险。区块链为特定交易创建标准合同，这些标准化合同将通过将其关键术语封装到区块链分布式分类账中，并在满足合同条款的情况下实现自动执行和结算，从而转化为智能合约。使用区块链设计中固有的自动托管功能，当合同的每个阶段执行时，将按照合同条款支付各方，从而消除了对可信第三方管理付款的需要。区块链作为一种去中心化的数据储存技术，其生成的数据具有不可篡改和不可伪造性，且任何一个节点都包含所有数据信息，数据极其安全。区块链技术正在提高交易的安全性，并为交易自动化打开了道路，因此，区块链技术目前逐渐被用来简化和提高气候融资中的交易流速度。

气候资金流动的障碍之一是许多交易规模相对较小，由于来源分散，且还要支付第三方中介的费用，交易成本较高，并且由于信息的不透明和不对称，捐助者和投资者对气候融资系统没有充足的信心，从而影响融资的进行。

区块链可以记录任何有价物的交易，使人员或组织之间的资金或资产转移透明化，增加投资者对气候融资的投资兴趣。同时，使用区块链不仅可以将较小的交易聚合，形成一笔大规模的交易，而且这种算法聚合和区块链的执行也可以用于联合投资者处理更大的交易，交易速度更快，成本更低。区块链技术还可以轻松地将承购商与项目进行匹配，通过地理位置、技术和商业运营等情况对双方进行分组。此外，区块链技术也可用于运行多方采购流程，包括需求方管理（即高系统负载时的减载）、项目再融资、环境债券和减排量购买协议（ERPA）。

目前，区块链在气候融资方面的应用已经小范围行动起来了，如CleanTek Market正在开发一个私有的完整区块链，以支持气候融资的交易服务。Gainforest正在使用智能合约来激励小规模的亚马孙农民保护热带雨林。

当遥感卫星验证某一特定的森林被保护成功时，智能合约便自动支付给农民一定的报酬。区块链技术使这些交易更加透明，并且可以被信任，由于没有中间人进行资金转移，行政成本下降明显。

当然，区块链技术在气候融资领域的应用也会面临跟其他领域一样的问题，目前只是处于起步阶段，尚待技术的完善使其与应用场景更好地匹配。

二、中国气候融资进展

2018年12月,在波兰卡托维兹举办的联合国气候变化框架公约第24届缔约方大会上,中国在"中国角"举办了一场主题为"中国气候投融资"的论坛,介绍了中国在气候融资方面作出的努力。近年来,中国不断发挥投融资对气候行动的杠杆和支撑作用,积极开展气候投融资活动,通过推动碳排放权交易市场、气候债券、气候保险、气候基金等金融机制和业务创新,为气候融资注入新的资金,撬动社会资本投入应对气候变化领域。

(一) 全国碳市场宣布启动,地方碳试点持续运行

1. 试点市场建设稳步推进

2011年10月,国家发展改革委批准北京、天津、上海、重庆、湖北、广东、深圳七个地区开展碳排放权交易试点工作,七个试点地区在2013—2014年陆续开始交易,之后市场建设稳步推进,2016年12月又新增了四川和福建两个试点地区。九个试点地区的运营为全国碳市场建设提供了示范和经验。

各个试点地区的政策设计,既有共性又因地制宜体现出各自的特点,相关政策的比较如表2-1所示。

二、中国气候融资进展

表 2-1

碳交易试点政策比较

试点碳市场	深圳	上海	北京	广东	天津	湖北	重庆	四川*	福建
交易平台	深圳碳排放权交易所	上海环境能源交易所	北京环境交易所	广州碳排放权交易所	天津碳排放权交易所	湖北碳排放权交易中心	重庆碳排放权交易中心	四川联合环境交易所	海峡股权交易中心
建立时间	2013.06.18	2013.11.26	2013.11.28	2013.12.19	2013.12.26	2014.04.12	2014.06.19	2016.12.16	2016.12.22
配额无偿分配	逐年分配	2013年三年分配；2016年后逐年分配	逐年分配	逐年分配	逐年分配	逐年分配	逐年分配	逐年分配	逐年分配
有偿分配	拍卖或固定价格出售	拍卖	拍卖	拍卖	拍卖或固定价格出售	拍卖，比例不超过政府预留配额的30%	—	拍卖	适时引入
政府预留配额	年度配额总量2%	—	年度配额总量5%	年度配额总量5%	—	年度配额总量10%	—	—	年度配额总量10%
CCER抵消比例和相关规定	不超过10%，本市内项目	不超过1%，2013年1月1日后，非水电类项目	不超过5%，本市内项目至少占50%	不超过10%，本市内项目占70%以上	不超过10%	不超过10%，本省内项目	不超过8%，本市内项目	—	不超过5%，本省内，非水电项目
除CCER外的其他抵消机制	—	—	林业碳汇	碳普惠核证自愿减排量（PHCER）	—	—	—	—	福建林业碳汇（FFCER）

续表

试点碳市场	深圳	上海	北京	广东	天津	湖北	重庆	四川*	福建
个人投资者	是	否	是	是	是	是	是	是	否
境外投资者	是	否	否	否	否	否	否	否	否
涨跌幅限制	10%（大宗交易为30%）	30%	公开交易20%	10%（挂牌竞价和挂牌点选）	10%	10%	20%	10%（大宗交易为30%）	挂牌点选10%；协议转让30%
远期产品	否	是	否	否	否	是	否	否	否

＊四川碳排放权交易市场目前暂无配额交易，故相应的纳管企业具体准入人要求、政府预留配额数量以及CCER抵消情况还未公布。

截至 2018 年底，各试点地区除四川省外，均已启动配额交易。四川省碳市场还在前期建设中，预计 2019 年启动配额模拟交易，2020 年开展配额现货交易。

试点碳市场的配额分配采用无偿分配为主，有偿分配为辅的方式，并逐步提高有偿分配的比例。有偿分配最常用的方式是企业竞价拍卖，在配额分配过程中，试点地区政府会预留一定比例的配额用于市场调节，以维持市场稳定。此外，各试点地区均设置抵消机制用于抵消部分配额，但对抵消比例的设置和抵消要求上各试点地区规定不同。抵消机制的设置扩大了碳市场的影响范围，使未加入碳市场的企业也能加入节能减排活动中，增加了市场活跃度。

目前各碳试点市场主要以现货交易为主，有一些试点地区进行了碳金融和融资工具方面的尝试，但产品数量不多，金额也不大。

2. 全国碳市场处于基础建设期

2017 年 12 月 19 日，中国宣布启动全国碳排放权交易体系，受到全球关注。国家发展改革委于 2017 年 12 月 18 日印发了《全国碳排放权交易市场建设方案（发电行业）》，明确了我国碳市场建设的指导思想和主要原则，对我国建设全国碳排放权市场具有重要指导意义。

该方案提出全国碳市场启动后，将经历三大建设阶段：一是基础建设期，为期一年左右，主要完成全国统一的数据报送系统、注册登记系统、交易系统建设以及碳市场管理制度建设；二是模拟运行期，为期一年左右，主要开展发电行业配额模拟交易，全面检验市场各要素环节的有效性和可靠性，强化市场风险预警与防控机制，完善碳市场管理制度与支撑体系；三是深化完善期，也就是全国碳市场开始交易运行的阶段，其间主要将在电力行业间进行碳配额现货交易，在电力行业稳定运行后，将扩大行业覆盖范围和交易品种，并尽早将中国核证自愿减排量（CCER）纳入碳市场。建设阶段如图 2－1 所示。

图 2-1 全国碳市场建设阶段图

全国碳市场的启动将以发电行业作为突破口，之所以这样选择，是因为发电行业历史数据较完整，产品单一，主要是热、电两类，管理较规范；且发电行业温室气体排放量大，根据方案的标准，发电行业纳入碳排放监管机制的企业达到1700余家，总体排放量将超30亿吨。据测算，仅纳入全国碳市场电力行业碳排放量的规模已超过目前第一大规模的欧盟碳排放交易体系（EU-ETS），中国碳市场全面启动后势必成为全球最大规模的碳排放权交易市场。

目前全国碳市场处于基础建设期。现阶段已确定全国市场的登记注册系统设置在湖北，交易结算系统设置在上海，分别由湖北碳排放权交易中心和上海环境能源交易所牵头做相关系统的基础建设工作。将全国碳市场不同系统建立在两个不同省市主要是考虑到两地的资源优势：湖北碳市场开户会员数、市场参与人数等有效指标均居全国第一，与各非试点地区的合作交流紧密，在会员整合上有较大优势；而上海中国金融中心的地位使其拥有良好金融市场环境，建立全国碳市场的配套服务较为完善。

3. 2018年各试点市场交易价格和交易量差距较大

根据各试点地区官网披露交易信息统计可得，2018年1月至12月，除四川外的八个试点地区累计成交量约6242万吨，累计成交金额约12.6亿元，其中各试点地区的交易量和交易价格差距较大：广东、深圳的成交量和成交额较大，广东为2836.20万吨和35346.15万元，深圳为1267.95万吨和29825.82万元；重庆、天津的成交量和成交额较小，重庆为26.94万吨和117.48万元，天津为228.78万吨和2654.84万元。在价格方面，北京配额的平均价格最高，约为57.93元/吨；重庆价格最低，约为4.36元/吨，如图2-2、表2-2所示。

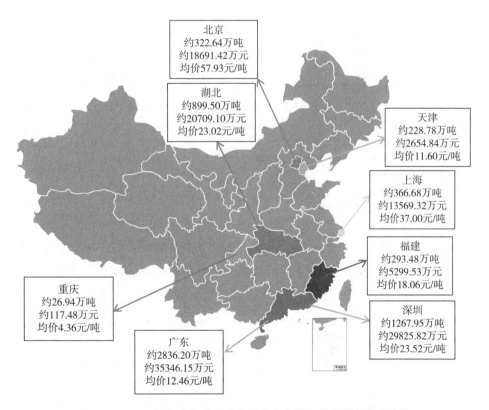

图2-2 2018年各试点地区碳交易总成交量、成交额和成交均价

表2-2   2018 各试点地区日均成交价格最高与最低价

地区	北京	深圳	上海	重庆	福建	湖北	广东	天津
日期	2018-10-09	2018-06-08	2018-07-30	2018-01-25	2018-11-20	2018-11-22	2018-12-28	2018-06-27
最高日均成交价格(元/吨)	74.60	58.91	42.58	31.93	30.00	32.71	18.87	13.96
日期	2018-09-20	2018-12-13	2018-01-10	2018-06-26	2018-12-21	2018-02-01	2018-10-09	2018-06-22
最低日均成交价格(元/吨)	30.32	5.24	27.79	2.24	12.00	10.91	1.27	9.24

数据来源：根据各试点交易所公布交易信息统计分析所得。

（二）绿色金融对气候融资的推动力不断显现

绿色金融作为推动世界经济绿色化转型的金融解决方案正在全球范围内兴起。有预测显示，在未来五年内，中国每年需要投入至少2万亿—4万亿元人民币（约3200亿—6400亿美元）资金来应对环境和气候变化问题[①]。绿色金融可以发挥调动各方资金投入绿色发展的积极作用。发展绿色金融对中国经济可起到稳增长、调结构的作用，目前已经上升为国家战略。各种绿色金融工具如绿色信贷、绿色债券、绿色保险、绿色基金发展迅速，从而动员和激励更多社会资本，同时有效抑制污染性投资，持续为应对气候变化提供长效资金。

1. 绿色信贷

绿色信贷是中国绿色金融体系中起步最早的领域。该理念最早于2007年由中国人民银行、银监会等机构在《关于落实环境保护政策法规防范信贷风险的意见》中提出。根据中国银监会统计，自2013年有统计数据以来，中国21家主要银行绿色信贷规模呈稳定增长态势，余额从2013年6月末的48527亿元人民币增至2017年6月末的82957亿元人民币[②]。其中，节能环保项目和服务贷款余额从34294亿元增至65313亿元，节能环保、新能源、新能源汽车等战略性新兴产业制造端贷款余额从14233亿元增至17644亿元（见图2-3、图2-4）。

[①] 详见: http://wemedia.ifeng.com/67666860/wemedia.shtml.

[②] 21家主要银行包括：国家开发银行、中国进出口银行、中国农业发展银行、中国工商银行、中国农业银行、中国银行、中国建设银行、交通银行、中信银行、中国光大银行、华夏银行、广东发展银行、平安银行、招商银行、浦东发展银行、兴业银行、民生银行、恒丰银行、浙商银行、渤海银行、中国邮政储蓄银行。详见银保监会官网: http://www.cbrc.gov.cn/chinese/files/2018/8FF0740BF974482CB442F31711B3ED03.pdf.

二、中国气候融资进展

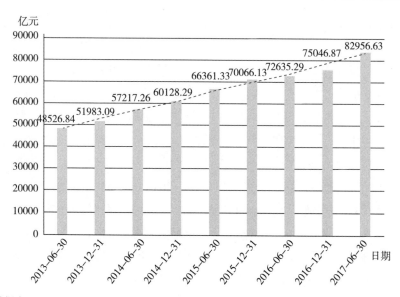

数据来源：银保监会官网，http：//www.cbrc.gov.cn/chinese/home/docView/96389F3E18E949D3A5B034A3F665F34E.html.

图 2－3 中国 21 家主要银行绿色信贷余额

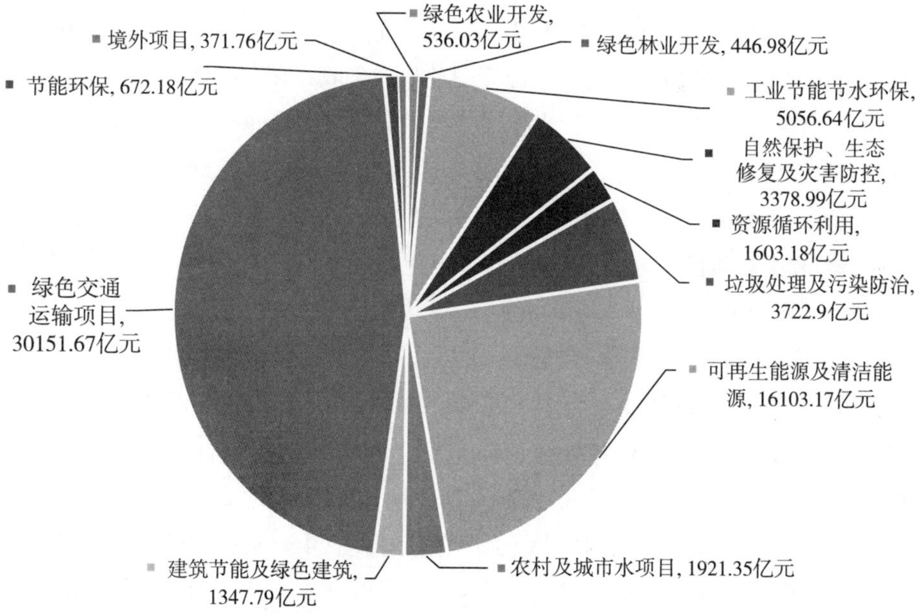

数据来源：银保监会官网，http：//www.cbrc.gov.cn/chinese/files/2018/8FF0740BF974482CB442F31711B3ED03.pdf.

图 2－4 中国 21 家主要银行绿色信贷余额（截至 2017 年 6 月末）

19

作为中国绿色投融资最重要的渠道之一，绿色信贷在撬动气候资金方面发挥了重要作用。绿色交通和清洁能源项目一直是绿色信贷资金投向的两个主要领域。2017年6月末，投向两者的贷款余额分别为30152亿元和16103亿元，预计总共可节约标准煤1.79亿吨，减少排放CO_2e 4.33亿吨。

中国近年来正大力激励和引导绿色信贷的发展。在政策推动方面，2017年12月，中国人民银行印发《关于推广信贷资产质押和央行内部（企业）评级工作的通知》，宣布优先接受符合标准的绿色贷款作为信贷政策支持再贷款、SLF等工具的合格信贷资产担保品。2018年1月，中国人民银行发布《关于建立绿色贷款专项统计制度的通知》，明确绿色贷款的标准，建立有效的绿色信贷考核评价体系；同年6月，中国人民银行宣布将优质绿色贷款纳入MLF担保品范围；随后人民银行于2018年7月27日发布了《中国人民银行关于开展银行业存款类金融机构绿色信贷业绩评价的通知》，并出台《银行业存款类金融机构绿色信贷业绩评价方案》，表明银行绿色信贷的执行情况将直接与宏观审慎评估结果挂钩。政策的出台可以刺激银行金融机构重点发展绿色信贷，对整个绿色信贷市场起到正面促进作用。

中国绿色信贷在产品形式上积极创新，从而吸引更多社会资本流向绿色和气候领域。产品创新不仅包括合同能源管理项目未来收益权、碳排放权、排污权等新兴绿色权益资产抵质押贷款，也包括直接投向具有二氧化碳减排效应的绿色项目。多家银行在2018年推出绿色信贷创新产品，建设银行于2018年6月率先推出绿色资产证券化项目，与5家企业签约共38亿元的项目，有力支持节能减排、清洁能源利用等项目，为减缓和适应气候变化领域释放资金。

2. 绿色债券

绿色债券近年来的发展速度令人瞩目。在利好政策刺激和有效金融监管的双重作用下，绿色债券在中国的规模不断扩大，体量和数量均保持强劲增长势头。2018年，中国境内外绿色债券发行规模达2675.93亿元，与2017年的2477.14亿元相比上升了8.02%，其中境内贴标绿色债券共发行129只，发行规模达2221.97亿元；境外贴标绿色债券发行15只，约合人民币453.96亿元。中国绿色债券总发行规模在全球占比23.27%，与2017年的24.59%相比有所下降，但依然是全球最大的绿色债券发行市场之一①。

---

① 中央财经大学绿色金融国际研究院. 中国绿色债券市场2018年度总结［R］. 2018.

2018年发行绿色债券的募集资金投向如图2-5所示①。除金融债外，清洁能源是绿色债券募集资金投向最多的领域，共计229.98亿元。其次是清洁交通（202.5亿元）和资源节约与循环利用（57.2亿元）领域。清洁能源项目主要涉及风力发电、太阳能光伏发电、水力发电等。

金融机构在债券创新方面做了许多有利尝试，例如向广大社会主体募集资金，用于投向更多气候改善领域。债券品种的创新，既扩大了绿色债券的应用领域，也增加了绿色债券的市场份额。

绿色债券创新不仅体现在品种创新上，在债券受众选择上也更加多元化。国家开发银行发行中国首只面向个人投资者的零售绿色债券，提高了社会大众的环保意识，形成了良好的社会效应。中国进出口银行发行国内首只面向全球投资者的绿色债券，有效连接了国内外资本市场，引进国外资金。另外，浙江泰隆商业银行发行国内首只绿色小微金融债，募集资金投向污染防治、资源节约与循环利用、清洁能源、生态保护和适应气候变化四个方面的绿色信贷项目，将绿色信贷与绿色债券进行结合。

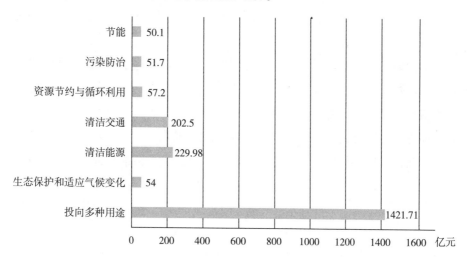

数据来源：中央财经大学绿色金融国际研究院．中国绿色债券市场2018年度总结［R］．2018.

图2-5　2018年中国绿色债券募集资金投向

---

① 募集资金投向根据中国金融学会绿色金融专业委员会发布的《绿色债券支持项目目录（2015年版）》资金用途六大类分类，由于金融债募集资金用途需根据后续信息披露情况进行确定，因此设置"投向多种用途"类别用于归类。

3. 绿色保险

绿色保险在利好政策的出台下逐渐发挥其在生态治理、绿色投资、风险管理方面的作用，是一项重要的气候风险分散机制。目前绿色保险正在国内各省市试点推行，各大保险公司也正在积极创新绿色保险险种。在新能源行业，已有光伏辐照指数保险、光伏组件效能保险、风电指数保险等新型绿色保险创新；在防范因极端天气导致的自然灾害方面，也有农业巨灾保险、气候指数保险等风险管理保险的尝试。

中国政府在绿色保险政策保障方面做了许多有益的尝试，通过健全制度、加快立法，助力绿色保险增强中国应对气候变化能力，防止气候灾害给利益相关方带来经济损失。2017年5月，中国保监会制定首个环保责任保险金融行业标准——《化学原料及化学制品制造业责任保险风险评估指引》，为保险公司承保前风险评估提供参考标准。2018年5月，生态环境部审议并通过了《环境污染强制责任保险管理办法（草案）》，进一步规范健全了环境污染强制责任保险制度。相关政策的出台，推动了绿色保险体系逐渐覆盖清洁能源、生态修复、环境治理、生态农业、绿色建筑和绿色交通等多个领域。

中国多省市均进行了绿色保险的创新尝试，险种包括环境污染强制责任险、绿色责任险、农业保险等。2017年，上海率先推出了全国首个农业台风巨灾指数保险，为上海、浙江等沿海八省市从事农业生产及相关产业的单位和个人提供台风巨灾保障。该巨灾指数保险在保险科技、保障范围、普惠属性三方面都进行了创新，结合了中国农业特色，实现精准的极端气候致灾风险评估[①]（见图2-6）。

---

① 详见：http://shanghai.circ.gov.cn/web/site7/tab359/info4077243.htm。

二、中国气候融资进展

图 2-6　绿色保险主要分类示意图

4. 绿色基金

财政部清洁发展机制基金（CDMF）作为发展中国家首只应对气候变化的政策基金，从 2010 年全面运行以来，通过社会性基金管理模式，截至 2017 年 12 月 31 日已累计投资绿色低碳项目 265 个，实现减排温室气体 5013.63 万吨 CO_2e[①]。而在地方层面，中国各地方政府发起绿色发展基金成为一种趋势，具有巨大的市场潜力。中国现有绿色基金 428 只，其中约有 90% 的基金为绿色产业基金。

绿色基金主要投向清洁能源领域，具体到细分领域，以新能源汽车和分布式能源行业为主。此外，环保行业中的污水处理和环境修复工程，也是绿色基金的重要投资领域。绿色基金对改善中国生态环境、减少温室气体排放、增强气候适应性等起到促进作用。

（三）PPP 模式在生态环保领域稳步推进

2017 年以来，国家稳步推进社会资本深入参与 PPP 项目，尤其是生态环保领域。在国家高度重视生态文明建设的政策引导之下，近几年生态建设和

---

① 中国清洁发展机制基金. 中国清洁发展机制基金 2017 年度报告 [R/OL]. China Clean Development Mechanism Fund 2017 Annual Report，2017. http：//www.cdmfund.org/zh/jjnb/20609.jhtml.

环境保护相关投资力度不断加大。2017年7月财政部发布的《关于政府参与的污水、垃圾处理项目全面实施PPP模式的通知》指出，将推进政府参与的新建污水、垃圾处理项目全面实施PPP模式，在该领域建立以社会资本为主，统一、规范、高效的PPP市场。

根据财政部PPP项目信息库数据，截至2018年12月31日，管理库项目共计8654个，投资额13.2万亿元；季度环比净增项目365个，投资额8822亿元；年度同比净增项目1517个，投资额2.4万亿元。目前，管理库中处于执行和移交阶段的项目（已落地项目）4691个（目前移交阶段项目0个），投资额7.2万亿元，落地率54.2%（即已落地项目数与管理库项目数的比值）。污染防治与绿色低碳领域共有项目4766个、投资额4.7万亿元，分别占管理库的55.1%和35.6%，年度同比新增项目787个，投资额0.6万亿元（见图2-7）。

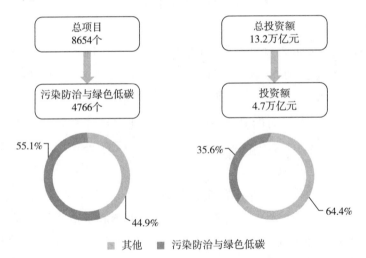

图2-7 2018年财政部PPP项目库总项目及污染防治与绿色低碳项目统计

2017年PPP项目改革稳中求进，在示范项目评选过程中更加注重项目实施的规范性和高质量。第三批PPP示范项目中，生态建设和环境保护类项目共计46个，以综合治理项目为主，投资总额810.56亿元，项目数量占总项目数的8.9%，投资总额占总投资额的6.9%。2017年，第四批示范项目共有396个，占全部申报项目的32%，投资总额逾7588.44亿元，其中生态建设和环境保护类示范项目共计37个，投资总额550.62亿元，项目数量占比

9.3%，投资总额占比7.3%。与第三批相比，由于第四批示范项目评选更注重项目质量，生态建设和环境保护类项目的投资金额和项目数量都有所下降，但由于项目总数也下降了23.3%，总投资额下降了35.1%，故生态建设和环境保护类项目投资金额和项目数量的占比都有所增加（见图2-8）。

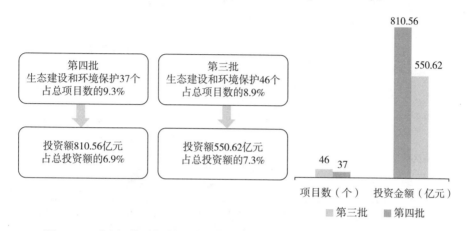

图2-8 财政部第三批和第四批PPP项目中生态建设与环境保护项目统计

2017年全国公共财政对节能环保资金支持比2016年有较大增长，达5617.33亿元，同比增长18.6%。其中，中央财政资金的支持为350.56亿元，与2016年相比增加55.07亿元；地方财政资金的支持为5266.77亿元，占地方财政总支出的比例增至3.04%。具体投向方面，全国公共财政用于污染防治与自然生态保护的决算分别为1883.02亿元和537.10亿元，同比增长分别达30.1%和64.5%；能源管理事务支出达到787.49亿元，较2016年增长54.2%。2017年，全国可再生能源的一般公共预算支出为52.99亿元，同比下降38.5%，其中主要是地方财政预算支出减少，中央在该领域的公共预算支出也有所下降（见图2-9、表2-3）。

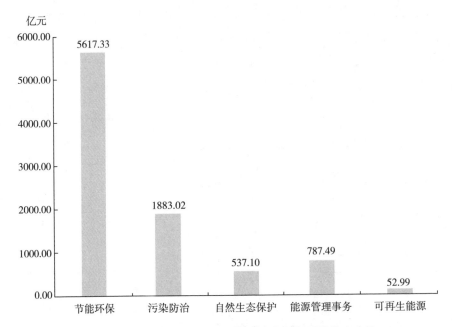

图 2 - 9 2017 年公共财政对气候变化相关领域的资金支持

表 2 - 3 中央财政和地方气候变化支出与其他项目的比较 单位：亿元

年份	中央			地方		
	2015	2016	2017	2015	2016	2017
节能环保	400.41	295.49	350.56	4402.48	4439.33	5266.77
可再生能源	2.10	10.88	8.71	162.61	75.24	44.28
教育	1358.17	1447.72	1548.39	24913.71	26625.06	28604.79
科学技术	2478.39	2686.10	2826.96	3384.18	3877.86	4440.02
文化体育与传媒	271.99	247.95	270.92	2804.65	2915.13	3121.01
医疗卫生	84.51	91.16	107.6	11868.67	13067.61	14343.03
总支出	80639.66	86804.55	94908.93	150335.62	160351.36	173228.34
节能环保占比（%）	0.5	0.34	0.37	2.93	2.77	3.04

数据来源：财政部预算司财政数据，http://yss.mof.gov.cn/2016js/.

二、中国气候融资进展

表2-4 PPP示范项目绿色低碳相关行业统计表

领域	第一批示范项目 项目数	第一批示范项目 投资额（亿元）	第二批示范项目 项目数	第二批示范项目 投资额（亿元）	第三批示范项目 项目数	第三批示范项目 投资额（亿元）	第四批示范项目 项目数	第四批示范项目 投资额（亿元）	合计 项目数	合计 投资额（亿元）	比例（%）
1.交通	9	1565.96	38	3491.78	40	4467.48	41	2375.99	128	11901.21	43.57
轨道交通	7	1526.51	13	2452.67					20	3979.18	14.57
高速公路			7	613.94	26	3689.36	19	1927.12	52	6230.42	22.81
非收费公路			8	232.38					8	232.38	0.85
交通枢纽			4	17.32	2	20.57	4	32.14	10	70.03	0.26
机场			2	81.98	1	203.17			3	285.15	1.04
铁路			2	45.7	3	126.41	1		6	172.11	0.63
公交			1	15					1	15.00	0.05
桥梁	2	39.45	1	32.78	4	370.09	2	33.02	7	435.89	1.60
其他					4	57.87			6	97.32	0.36
2.市政工程	8	77.27	66	810.53	119	1495.35	163	1999.27	356	4382.42	16.04
垃圾处理	1	5.26	22	97.24	31	124.44	21	57.6	75	284.54	1.04
地下综合管廊	1	13	14	407.77	31	838.68	46		46	1259.45	4.61
公园			1	25	4	43.94	3	31.82	8	100.76	0.37
供气			2	3.64	2	3.4	1	4.22	5	11.26	0.04
供热	3	26.4	6	59.52	13	62.24	13	66.87	35	215.03	0.79
供水	3	32.61	18	190.11	24	152.65	12	91.89	57	467.26	1.71

27

续表

领域	第一批示范项目		第二批示范项目		第三批示范项目		第四批示范项目		合计			
	项目数	投资额（亿元）	项目数	投资额（亿元）	项目数	投资额（亿元）	项目数	投资额（亿元）	项目数	比例（%）	投资额（亿元）	比例（%）
海绵城市			1	13.85	5	208.9	14	222.55	20	1.74	445.30	1.63
绿化			1	5.5	4	16.22	2	14.13	7	0.61	35.85	0.13
排水			1	7.9	5	44.88	6	31.62	12	1.05	84.40	0.31
3. 环境保护	11	111.99	31	742.46	82	932.79	77	922.86	201	17.51	2710.10	9.92
污水处理	9	59.08	15	337.49	40	181.37	36	508.63	100	8.71	1086.57	3.98
环境综合治理	2	52.91	15	364.97	38	711.32	29	413.66	84	7.32	1542.86	5.65
湿地保护			1	40	4	40.1	2	0.57	7	0.61	80.67	0.30
合计	28	1755.22	135	5044.77	241	6895.62	281	5298.12	685	59.67	18993.73	69.53

数据来源：财政部政府和社会资本合作中心PPP项目库。

(四) 中国积极参加国际交流合作，为全球应对气候变化作贡献

中国在国际上持续支持全球气候治理，促进、引导气候变化全球性合作，积极支持发展中国家应对气候变化，为最不发达国家、非洲国家及其他发展中国家提供了实物及设备援助，对其参与气候变化国际谈判、政策规划、人员培训等方面提供大力支持，并在发展中国家启动开展10个低碳示范区、100个减缓和适应气候变化项目及1000个应对气候变化培训名额的合作项目①。

1. 积极推进联合国框架下的气候谈判进程

近年来，中国深度参与《巴黎协定》后续谈判，推动建立新的全球气候治理体系。2018年12月在波兰卡托维兹举办的第24届联合国气候变化大会上，中国积极参与谈判过程，推动各方代表就关键问题达成一致，为最终落实《巴黎协定》实施细则作出建设性贡献。此外，中国在"中国角"举办"中国气候投融资"论坛，介绍了中国在气候融资方面所取得的成果并分享了成功经验。

2. 深度参与公约外气候变化相关事务

中国在彼得斯堡气候变化对话以及二十国集团会议、蒙特利尔议定书、国际民航、国际海事等组织的气候变化相关议题中发声，并持续关注联合国大会、亚太经合组织、金砖国家会议等场合下气候变化相关问题。2018年6月，中国与欧盟、加拿大在比利时布鲁塞尔共同举办第二次气候行动部长级会议，在全球应对气候变化进程不确定性增强的背景下进一步凝聚各方共识，注入新的政治推动力。2018年9月，中国作为17个发起国之一与其他国家共同设立全球适应委员会（GCA），推动国际社会适应气候变化通力合作，加速全球气候行动进程，帮助气候脆弱型国家提高气候适应力②。

3. 致力推动气候变化"南南合作"

多年来，中国致力于推动气候变化"南南合作"。截至2018年4月，中

---

① 详见：http://www.china.com.cn/news/2017-10/31/content_41820656.htm.
② 中国生态环境部. 中国应对气候变化的政策与行动2018年度报告［R］. 2018.

国已与 30 个发展中国家签署合作谅解备忘录,赠送遥感微小卫星、节能灯具、户用太阳能发电系统帮助其应对气候变化。中国对 80 多个发展中国家进行清洁能源、低碳示范、农业抗旱技术、水资源利用和管理、智能电网、绿色港口、水土保持、紧急救灾等领域技术援助,通过开展减缓和适应气候变化项目、赠送节能低碳物资和监测预警设备、组织应对气候变化南南合作培训等方式提升其他发展中国家应对气候变化的能力①。

4. 积极推动南北交流合作

中国广泛参与各国间及国际组织的交流活动,积极组织参与国际会议,深化与世界银行、亚洲开发银行、联合国开发计划署等多边机构的合作。中国与新西兰、德国、法国、加拿大等多国举行了气候变化双边合作机制会议,与美国、欧盟、法国、德国、英国、加拿大、日本等国家和地区在碳市场、低碳城市、适应气候变化等领域开展交流合作。

---

① 中国生态环境部. 中国应对气候变化的政策与行动 2018 年度报告 [R]. 2018.

表2-5 中国气候融资发展情况

分类		规模	时期（年）	说明	数据来源
中国碳市场	中国碳市场试点	2018年全年试点碳市场配额成交量达6200万吨余吨，成交额约12.6亿元	2018	2018年1月至12月，除四川外的八个试点地区累计成交量约6242万吨，累计成交额约12.6亿元，其中各试点的交易量和交易额价格差距较大：广东、深圳的成交量和成交额较大，重庆、天津的成交量和成交额较小；在价格方面，北京配额平均价格最高，重庆价格最低	课题组根据碳交易试点网站公开资料整理
慈善资金	中国绿化基金会所获赠款	4257万元	2017	2017年全年公益事业支出4541万元	《中国绿化基金会2017年度审计报告》
	中国所接收国内外款物捐赠中流向生态环境领域的数额	20.7亿元	2017	2017年全年生态环境领域共接受捐赠20.7亿元，占总捐赠额的1.38%，较2016年增长35.3%	《2017年度中国慈善捐助报告》
传统金融市场	传统国际金融市场	中国对可再生能源的投资	1266亿美元	2017	中国的可再生能源投资规模庞大，2017年同比增长30.7%	《2018年可再生能源全球现状报告》
	国内金融市场	中国主要银行业金融机构绿色信贷余额	8.22万亿元	2017.06	—	Wind资讯

二、中国气候融资进展

31

续表

分类		规模	时期（年）	说明	数据来源
传统金融市场	国内金融市场	中国银行业节能环保贷款余额	6.53 万亿元	2017.06	—	Wind 资讯
		中国银行业战略性新兴产业贷款余额	1.69 万亿元	2017.06	—	Wind 资讯
		国内财政资金	6353 亿元	2018	—	财政部网站统计数据
		国内公共财政中节能环保投入	2675.93 亿元	2018	—	财政部网站统计数据
		贴标绿色债券规模总量	16150.70 亿元	2018	—	《中国绿色债券市场2018年度总结》
		非贴标绿色债券发行规模总量		2018	—	
	国内清洁技术领域	中国清洁技术行业获得 VC/PE 投资规模	12.41 亿美元	2018	2018 年，清洁技术行业共发生50起融资案例，与 2017 年相比较大幅下跌 20.63%；年度清洁技术行业融资规模为 12.41 亿美元，与 2017 年相比大幅下降 64.81%	投中集团旗下金融数据产品 CVSource
		中国清洁技术市场完成并购交易规模	79.69 亿美元	2018	2018 年清洁技术行业完成并购案例数量与 2017 年相比减少 29.23%，跌至 2013 年以来的最低水平；完成并购交易规模为 79.69 亿美元，与 2017 年相比上升 17.3%	投中集团旗下金融数据产品 CVSource

续表

	分类		规模	时期（年）	说明	数据来源
企业直接投资	国际市场和国内市场	中国清洁技术企业境内IPO融资	11.52亿美元	2018	2018年度，清洁技术行业共有7家中企成功上市，主要集中在环保节能领域，IPO企业数量与2017年相比下滑41.67%。2018年度，清洁技术行业募资规模为11.52亿美元，与2017年相比下跌27.04%	投中集团旗下金融数据产品CVSource
PPP项目	入库项目	生态建设和环境保护	9080亿元	2018	—	财政部PPP项目库
		林业	765亿元	2018	—	
	对外发布项目	污染防治与绿色低碳项目	4.7万亿元	2018	截至2018年12月31日，管理库中污染防治与绿色低碳项目4766个，投资额4.7万亿元；年度同比净增项目787个，投资额0.6万亿元	

三、"一带一路"沿线的气候投融资

应对气候变化是一个长期的有利于人类可持续发展的重大课题,单靠一个国家无法有效解决,需要全球通力合作。近几年,中国作为最大的发展中国家,不仅自身积极地应对气候变化,也积极地对外提供气候援助。2013年9月和10月,中国国家主席习近平先后提出了"丝绸之路经济带"和"21世纪海上丝绸之路"这两个符合亚欧经济整合的经济发展战略。"一带一路"沿线国家和地区人类活动比较集中,且相当多的国家生态环境脆弱,对气候变化适应能力较弱,亟须气候资金投入以改善生态环境,发展可持续经济。因此,建设绿色"一带一路"是"一带一路"倡议顶层设计中的重要内容。

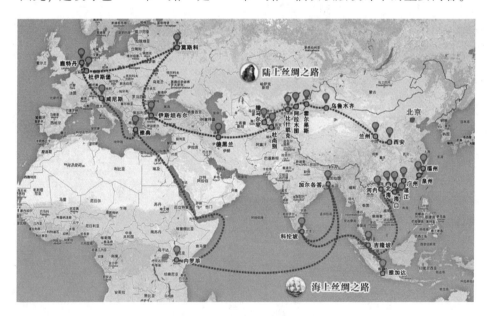

图3-1 "一带一路"示意图

三、"一带一路"沿线的气候投融资

（一）"一带一路"倡议的基本情况

1. "一带一路"倡议的提出顺应世界发展潮流

2008年国际金融危机之后，世界经济复苏进程减缓，金融危机的深层次影响逐渐显现，国际投资贸易格局有待调整，各国均面临并不乐观的发展形势。在此世界发展大背景下，2013年中国提出"一带一路"倡议，包括"丝绸之路经济带"和"21世纪海上丝绸之路"，它是一种合作发展的新理念和倡议。"一带一路"倡议恪守联合国宪章的宗旨和原则，秉持"共商共建共享"的原则，全面推进政策沟通、设施联通、贸易畅通、资金融通、民心相通的"五通"原则。"一带一路"倡议的提出顺应世界多极化、经济全球化、文化多样性、社会信息化的发展趋势，有利于建立全球自由贸易体系和开放型世界经济[①]。

"一带一路"贯穿亚欧非大陆，东起东亚经济圈，西至欧洲经济圈，包含了六条国际经济合作走廊，即新亚欧大陆桥、中蒙俄经济走廊、中国－中亚－西亚经济走廊、中国－中南半岛经济走廊、中巴经济走廊、孟中印缅经济合作走廊[②]，为应对气候变化的全球性合作提供了绝佳平台。在世界经济一体化、经济发展可持续化等特征下的新经济发展时期，构建互相合作与共同发展的经济大走廊，将给中国以及"一带一路"沿线国家和地区带来更广阔的发展机会，提升国家经济实力与竞争力。

2. "一带一路"建设取得阶段性进展

"一带一路"倡议自2013年提出以来，已经取得突破性进展。截至2018年上半年，中国已与103个国家和国际组织签署了共建"一带一路"合作文件，对沿线国家直接投资超过700亿美元，年均增长7.2%，在沿线国家新签对外承包工程合同额超过5000亿美元，年均增长19.2%[③]。商务部数据显示，截至2018年上半年，中国与24个"一带一路"沿线国家在建境外经贸合作区82家，新增投资

---

① 国家发展改革委，外交部，商务部. 推动共建"丝绸之路经济带"和"21世纪海上丝绸之路"的愿景与行动[J]. 智富时代，2015（3）：82-87.

② 刘宗义. 我国"一带一路"倡议在东南、西南周边的进展现状、问题及对策[J]. 印度洋经济体研究，2015（4）：92-109.

③ 详见：https://www.sohu.com/a/250863721_498798.

25.9 亿美元，占中国境外经贸合作区新增总投资的 87%。

（二）"一带一路"沿线主要国家的气候特征

"一带一路"倡议辐射地区包括中亚、西亚、南亚、中东、中南亚、北非、东非、中东欧等地的 65 个国家和地区，涉及沿线 44 亿人口，占世界总人口数的 63%，GDP 规模达 21 万亿美元，占世界经济总量的 29%，涵盖了世界上经济最具活力和潜力的大部分地区[①]。

1. 生态环境脆弱

"一带一路"沿线国家生态环境受气候变化影响较大，环境承载能力较差。沿线中亚区域各国多是沙漠、荒漠地区，绿色植被非常少，水资源匮乏，环境承载力十分脆弱；而东南亚地区的环境压力也在加剧，高速的商业开发和工业化水平的急速扩张，使热带雨林的面积快速缩小，各种工业污染也日益加剧。而像西亚、北非、东南亚和南亚等国，沿岸工厂大量向海洋排污，使海洋生态系统遭到破坏，损害生物资源，造成海洋污染。沿线国家脆弱的生态环境难以承受高污染、高排放的投资[②]。

2. 气候风险高

从气候变化的物理影响和气候安全威胁两个方面对"一带一路"沿线的东南亚、南亚、中亚、东北非、中欧五个地区进行气候变化分析，可得到如下结论：

气候变化对这五个地区都有不小的影响：对东南亚地区的主要影响是导致海平面上升，沿海土地减少，粮食产量下降，进而导致贫困问题加剧，社会不稳定因素增加；对南亚地区的主要影响是淡水资源匮乏，粮食产量降低，贫困问题加剧，极端天气频发，引发边界冲突，致使产生居民安全问题；对中亚地区的主要影响是水资源短缺，导致农业生产困难，潜在气候移民数量加剧，且中亚地区植被对降水特别敏感，干旱天气造成生物多样性降低；对东北非地区的主要影响是粮食产量下降，导致因抢夺资源引发的暴力冲突事件不断升级，

---

① 丁俊发. "一带一路"须打通"五流"[J]. 中国储运，2016（10）：46.
② 杨振，申恩威. "一带一路"战略下加快沿线国家绿色投资的探讨[J]. 对外经贸实务，2016（9）：21-24.

极端主义和恐怖主义蔓延；对中欧地区的主要影响是水资源短缺和极端天气频发，影响居民安全以及产生资源争夺威胁①。

图 3-2　"一带一路"沿线主要气候风险示意图

3. 排放总量巨大，减排潜力可观

中国是世界上最大的二氧化碳排放国，很多"一带一路"沿线国家位于生态脆弱敏感区，尤其是亚洲国家，已成为世界上化石能源消耗增长最快的国家。从表 3-1 可以看出，2015—2017 年"一带一路"沿线国家碳排放整体呈上升趋势，且"一带一路"区域涵盖全球碳排放大国，如中国、印度、俄罗斯等，碳排放总量巨大。2017 年"一带一路"沿线国家碳排放总量约为 227.58 亿吨 CO_2e，占世界碳排放总量的 63%。这些国家经济发展模式较为粗放，生产经营活动给当地生态环境带来巨大负担，环境问题突出，污染严重，亟须调整经济发展模式，进行低碳投资。如采取有效措施，沿线国家未来存在较大的减排潜力。

---

① 王志芳. 中国建设"一带一路"面临的气候安全风险 [J]. 国际政治研究，2015，36（4）：56-72.

表 3–1　"一带一路"沿线主要国家 2015—2017 年碳排放总量表

单位：百万吨 CO_2e

区域	国家	2015 年	2016 年	2017 年
东亚 12 国	中国	9975.3554	9967.9553	10110.279
	马来西亚	245.0976	251.0896	254.5759
	新加坡	60.7206	62.8291	64.7651
	越南	184.5019	197.8016	198.8265
	缅甸	23.4407	24.0176	25.3332
	柬埔寨	6.9765	7.482	7.9383
	老挝	1.745	1.8985	1.958
	印度尼西亚	459.6288	464.8568	486.8438
	文莱	10.3553	9.9647	10.2269
	菲律宾	113.8232	119.2737	127.608
	蒙古国	20.3564	28.681	30.3907
	泰国	320.574	324.8341	330.8396
南亚 8 国	巴基斯坦	172.09	187.4054	198.81
	阿富汗	10.1074	12.2579	13.0147
	孟加拉国	83.5008	85.2536	88.0575
	斯里兰卡	20.7134	21.9507	23.1384
	马尔代夫	1.4554	1.4721	1.5718
	尼泊尔	7.5854	8.526	9.0282
	印度	2276.4072	2377.4479	2466.7654
	不丹	0.98796	1.1146	1.1664
西亚 18 国	沙特阿拉伯	620.883	631.5492	635.0111
	伊拉克	167.7119	184.9387	194.4546
	伊朗	630.3618	637.5622	672.3123
	阿曼	64.2211	64.4421	65.1864
	卡塔尔	120.5472	118.6898	129.8033
	科威特	102.327	103.456	104.3935
	黎巴嫩	21.2692	19.7574	19.5468
	土耳其	380.8581	402.8208	447.8972
	埃及	200.4222	208.9113	218.6644
	叙利亚	28.9252	27.9994	27.9145

续表

区域	国家	2015年	2016年	2017年
西亚18国	约旦	23.1678	21.6301	21.3567
	以色列	66.0959	65.1682	66.5534
	也门	20.3846	19.1354	18.9658
	阿拉伯联合酋长国	227.4772	233.4265	231.7735
	巴林	32.8469	34.2651	34.4557
	塞浦路斯	6.9029	7.3078	7.5147
	希腊	74.9625	71.3731	76.0004
	巴勒斯坦	2.5541	2.3405	2.318
中亚5国	土库曼斯坦	75.379	76.5633	72.7025
	哈萨克斯坦	263.4839	278.3619	292.5885
	吉尔吉斯斯坦	9.9294	9.8599	10.4331
	塔吉克斯坦	5.4602	5.579	5.7114
	乌兹别克斯坦	110.6693	97.9253	98.9989
独联体7国	阿塞拜疆	40.2414	39.6883	38.2097
	亚美尼亚	5.7917	5.9608	5.6564
	白俄罗斯	58.9658	60.5092	61.3718
	摩尔多瓦	4.9487	5.2004	5.0936
	乌克兰	223.5797	235.156	212.1163
	格鲁吉亚	9.4636	9.8389	10.9517
	俄罗斯	1671.8951	1668.0699	1692.7948
中东欧16国	立陶宛	13.1416	13.1575	13.3939
	拉脱维亚	7.3337	7.2636	7.1668
	爱沙尼亚	15.8911	17.4935	19.8093
	波黑	24.7642	26.0017	26.6469
	阿尔巴尼亚	5.9477	6.2729	6.3792
	斯洛文尼亚	13.5991	14.3998	14.6096
	克罗地亚	17.9966	18.2215	17.1816
	塞尔维亚	41.3375	43.351	45.0928
	马其顿	7.2575	7.144	7.2515
	捷克	104.7648	106.5433	107.8958
	斯洛伐克	33.8975	33.9968	35.3862

续表

区域	国家	2015 年	2016 年	2017 年
中东欧 16 国	保加利亚	48.1327	45.2874	49.0712
	匈牙利	46.6652	47.5782	50.3447
	波兰	310.6151	322.234	326.6045
	罗马尼亚	77.7882	75.0517	79.9955
	黑山	2.4392	2.5735	2.6278
总计		22053.74086	22306.1691	22758.3457
全球		35462.7467	35675.0994	36153.2616

注：（1）表格所列国家名单根据新华丝路网所列国家名单以及各大新闻网站报道国家名单汇总。
（2）数据来源：Global Carbon Atlas, http://www.globalcarbonatlas.org/en/CO2-emissions。

（三）推动"一带一路"沿线绿色投资的基础

"一带一路"沿线多数国家的经济基础薄弱，环境问题突出，在经济建设过程中，生态环境的改善是一个很重要的考量因素。而沿线国家对外开放程度低、社会经济发展水平低，缺少气候投资理念。中国在"一带一路"建设中积极承担大国责任，引导中方企业在当地进行绿色投资，提出了建立绿色战略联盟，将环境治理与气候改善的相关经验分享给参与"一带一路"建设的相关方，在投资中保持经济、环境和社会效益的平衡。

1. 强化绿色理念："一带一路"绿色发展战略框架

针对"一带一路"国内国际建设过程中遇到的问题，2016 年环保部（2018 年改为生态环境部）提出"一带一路"绿色发展战略框架。该框架的提出是为了在"一带一路"建设中更为绿色化地进行产业输出、投资建设、技援合作，强化生态环保合作，规避生态环境风险。"一带一路"沿线涉及的国家人口众多，温室气体排放量巨大，如果依然采取粗放型的经济发展模式，《巴黎协定》设定的目标将难以实现。

该战略框架从四个层面进行设计：一是强化战略实施的顶层设计，实施绿色发展规划，特别是做好生态环保专项规划；二是要建立生态环境风险预警机制，精准识别和规避与"一带一路"沿线国家合作的生态环境风险；三是加强对外投资合作行为调控，激励对外投资；四是加强"一带一路"绿色发展战略实施的共通保障能力建设，建立完善的民间交流机制、详尽的宣传推广机制、发达的网络信息平台，做好绿色发展战略的实施支撑。具体战略内容见图3-3。

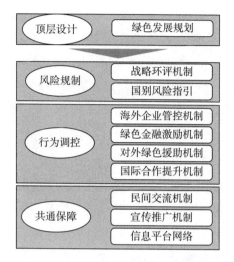

图3-3　"一带一路"绿色发展战略框架①

2. 加强绿色国际合作："一带一路"绿色发展国际联盟

构建绿色金融体系，实现金融体系绿色化，发展绿色经济，不仅是全球经济发展的新趋势，也是中国加强生态环保、生态文明建设的要求，更是未来"一带一路"建设的重要发展方向。

2017年5月14日，习近平主席在出席"一带一路"国际合作高峰论坛时提出建立"一带一路"绿色发展国际联盟，加强"一带一路"沿线各国创新合作，启动"一带一路"科技创新行动计划，开展科技人文交流、共建联合实验室、科技园区合作、技术转移四项行动，这一提议受到联合国以及社会各方的高度关注。该联盟致力于借助国际层面平台，促进"一带一路"沿线国家在环境保护、生态建设以及金融发展方面的合作。联盟的建立有助于加强新形势下的"南南合作""南北合作"、多边合作和多方合作，将吸引更多的商业机构、绿色基金、气候基金、非政府组织等团体进行环境保护方面的资金和技术援助②。目前，联合国环境规划署和中国生态环境部正在牵头

---

① 董战峰，葛察忠，王金南等．"一带一路"绿色发展的战略实施框架［J］．中国环境管理，2016，8（2）：31-35.

② 详见：http://jjckb.xinhuanet.com/2017-07/03/c_136412229.htm.

建立该联盟,并联合国际组织、企业、金融机构、智库及民间组织,尽快启动联盟工作。

3. 分享绿色治理经验:"一带一路"沙漠绿色经济创新中心

2017年6月24日,联合国环境规划署与亿利公益基金会在内蒙古鄂尔多斯市库布齐沙漠共同成立启动了"一带一路"沙漠绿色经济创新中心。该中心的创立旨在搭建一个交流平台,将库布齐30余年的沙漠绿色经济成功实践经验分享至全球荒漠化地区,推广库布齐治沙经验、成果与技术,这是推动"一带一路"绿色发展国际合作与科技创新的重要机制。

该中心下设三个业务平台,分别是"一带一路"生态技术研发平台、技术成果转移与应用平台和青年环境教育平台。该中心的工作内容是:开展科技创新推动沿线生态修复基础研究;培养生态修复与环境治理专业人才;举办环境保护培训和拓展活动增强青少年生态环保意识;开展"一带一路"沿线国家生态修复与绿色发展多边与双边合作;促进沿线国家生态修复产学研结合与科技成果产业化;分享荒漠化治理的中国经验,促进"一带一路"沿线各国生态治理与防治荒漠化国际合作①。

4. 凝聚绿色投资共识:"一带一路"绿色投资原则

为推动"一带一路"投资绿色化,2018年11月30日,中国金融学会绿色金融专业委员会与"伦敦金融城绿色金融倡议"共同发布了《"一带一路"绿色投资原则》。该原则在现有责任投资原则基础上,强化低碳与可持续发展理念,提高投资环境和社会风险管理水平。

为确保投资项目对气候和环境产生正外部性影响,支持《巴黎协定》的落实,该原则有以下七条倡议,包括将可持续性纳入公司治理,充分了解ESG风险,充分披露环境信息,加强与利益相关方沟通,充分运用绿色金融工具,采用绿色供应链管理,通过多方合作进行能力建设。

"一带一路"建设蕴含巨大气候投资机会。绿色治理平台的建立可为沿线各国建立应对环境问题共同体,提升环境风险应对能力。"一带一路"建设如果能够通过绿色输出、低碳发展的新理念,帮助沿线国家发展低碳经济,不仅可体现"一带一路"倡议平等互利、合作共赢、共同发展的宗旨,还能

---

① 详见:http://www.xinhuanet.com/energy/2017-06/26/c_1121212487.htm。

（四）中国对"一带一路"沿线国家的气候投融资项目

"一带一路"倡议提出以来，中方企业积极投入"一带一路"沿线国家项目建设，提升其减缓和适应气候变化的能力。"一带一路"沿线气候投融资呈现项目体量大、参与机构多、投资方式可选择性大的特点。

因"一带一路"沿线国家气候风险较高，社会经济发展水平低，基础设施建设落后，中国对其投资主要集中在清洁能源、清洁交通等基础设施上。一方面，这些投资能发挥基础设施的"乘数效应"，解决沿线国家迫切需求的持续稳定发展问题，改善人民整体生活水平；另一方面，基础设施的环境效益巨大，例如中国投资建设的巴基斯坦最大的水电站项目尼鲁姆—杰卢姆，该项目电站机组全部发电后，年发电量约为51.5亿千瓦时，占巴基斯坦水电发电量的12%，能解决巴基斯坦全国15%人口的用电紧缺问题，减排效应明显[①]。

截至2018年10月底，中国投资的"一带一路"沿线国家清洁交通及清洁能源项目有51个，其中清洁交通类项目20个，清洁能源类项目31个（见表3-2）。项目遍布东南亚、南亚、西亚、中亚、欧洲、非洲等地区。这些项目不仅可以满足当地居民的需求，提高居民整体生活水平，还可改善当地的气候条件。

> **案例1**
>
> **中国承建东非首条轻轨**
>
> 2015年9月20日，由中国承建的东非首条轻轨在埃塞俄比亚首都亚的斯亚贝巴通车试运行。亚的斯亚贝巴轻轨项目全长32公里，包括一号线和二号线两条线路，分别贯穿城市东西和南北。该轻轨每天运行16小时，每天有244班轻轨往返于亚的斯亚贝巴城中，日均载客量约10万人次，轻轨的运行将有力缓解亚的斯亚贝巴的交通拥堵。

---

① 详见：https://www.yidaiyilu.gov.cn/xwzx/hwxw/52755.htm.

该项目共投资 4.75 亿美元，其中 85% 由中国进出口银行提供贷款，15% 由当地政府自筹。目前运营的轻轨是埃塞俄比亚整个轻轨项目的第一工程阶段，预计能减轻 10% 的城市交通压力，该项目的第二阶段修建目前正在等待后续资金的投入。

轻轨项目的运行将有效减少当地的碳排放。在轻轨未投入运行前，城市交通排放量占亚的斯亚贝巴城内总温室气体排放量的大部分。根据 2012 年亚的斯亚贝巴温室气体排放清单显示，城中排放的 480 万吨 CO_2e 中，47% 来自交通运输车辆。自 2015 年下半年通车至 2018 年 5 月 31 日，轻轨已安全运营 985 天，累计行车 23 万列次，运营里程达 584.2 万公里，共计运送旅客 1.29 亿人次，日均客流 10.45 万人次，历史最高单日客流 18.5 万人次①。轻轨的运行有效减少了亚的斯亚贝巴的温室气体排放，根据彭博慈善事业和中国绿色技术开发商比亚迪的预测，预计到 2030 年，亚的斯亚贝巴轻轨项目将累计为该城市减排 180 万吨 CO_2e。2016 年，亚的斯亚贝巴被授予 2016 年 C40 城市奖，在全球范围内展示了其在应对气候变化方面的努力，而轻轨项目是其能够赢得此奖项的重要原因，该奖项肯定了亚的斯亚贝巴从柴油燃料公共交通向更生态友好型交通运输的转型。

表 3-2　　中国在"一带一路"沿线的气候投融资项目

序号	项目类别	项目所在国	项目名称
1	铁路交通	老挝	中老铁路项目
2	铁路交通	肯尼亚	肯尼亚蒙巴萨至内罗毕铁路项目
3	铁路交通	白俄罗斯	白俄罗斯铁路电气化改造项目
4	铁路交通	俄罗斯	莫斯科—喀山高铁项目
5	铁路交通	缅甸	中缅木姐—曼德勒铁路项目
6	铁路交通	缅甸	中缅国际铁路项目
7	铁路交通	孟加拉国	孟加拉国达卡至吉大港高速铁路项目
8	铁路交通	泰国	中泰铁路
9	铁路交通	泰国	曼谷大众捷运项目（粉红线和黄线）

---

① 详见：http://www.sohu.com/a/233930118_157267.

续表

序号	项目类别	项目所在国	项目名称
10	铁路交通	匈牙利、塞尔维亚	匈塞铁路
11	铁路交通	印度尼西亚	雅万高铁项目
12	铁路交通	埃塞俄比亚、吉布提	亚的斯亚贝巴—吉布提铁路
13	铁路交通	巴基斯坦	巴基斯坦拉合尔轨道交通橙线项目
14	铁路交通	沙特阿拉伯	麦加—麦地那高速铁路项目
15	城市轨道交通	以色列	红线轻轨项目
16	城市轨道交通	格鲁吉亚	格鲁吉亚现代化铁路项目
17	城市轨道交通	埃及	埃及斋月十日城市郊铁路项目
18	城市轨道交通	孟加拉国	达卡城市轨道交通项目
19	城市轨道交通	印度	孟买地铁1号线
20	水路交通	孟加拉国	帕德玛大桥及河道疏浚项目
21	风力发电	埃塞俄比亚	埃塞俄比亚阿达玛风电项目
22	风力发电	马耳他	黑山莫祖拉风电项目
23	风力发电	巴基斯坦	吉姆普尔风电项目
24	太阳能光伏发电	巴基斯坦	巴基斯坦旁遮普省900兆瓦光伏地面电站项目
25	太阳能光伏发电	巴基斯坦	巴基斯坦100兆瓦大型太阳能光伏电站
26	太阳能光伏发电	阿尔及利亚	阿尔及利亚233光伏电站项目
27	太阳能光伏发电	厄立特里亚	中国援厄立特里亚太阳能光伏项目
28	太阳能光伏发电	匈牙利	格林斯乐太阳能电站
29	太阳能光伏发电	阿根廷	高查瑞光伏电站
30	太阳能光伏发电	乌克兰	无
31	太阳能热+太阳能光伏发电综合项目	摩洛哥	努奥瓦尔扎扎特光热发电综合体
32	天然气项目	缅甸	仰光达吉达天然气联合循环电厂项目
33	水电项目	老挝	老挝南俄3水电站项目
34	水电项目	老挝	老挝胡埃兰潘格雷河水电站项目
35	水电项目	老挝	老挝南涧水电站
36	水电项目	巴基斯坦	巴基斯坦卡洛特水电站项目
37	水电项目	巴基斯坦	尼鲁姆—杰卢姆水电站
38	水电项目	科特迪瓦	苏布雷水电站项目
39	水电项目	尼泊尔	上马相迪A水电站

续表

序号	项目类别	项目所在国	项目名称
40	水电项目	埃塞俄比亚	吉布3水电站
41	水电项目	埃塞俄比亚	泰克泽水电站
42	水电项目	安哥拉	卡古路·卡巴萨水电站项目
43	水电项目	厄瓜多尔	厄瓜多尔科卡科多—辛克雷水电站
44	水电项目	柬埔寨	柬埔寨额勒赛水电站
45	水电项目	乌干达	卡鲁玛水电站
46	水电项目	几内亚	几内亚凯乐塔水电站
47	水电项目	喀麦隆	喀麦隆曼维莱水电站
48	水电项目	马来西亚	马来西亚巴勒水电站
49	水电项目	白俄罗斯	维捷布斯克水电站
50	地热项目	印度尼西亚	印度尼西亚 SMGP 240MW 地热发电项目
51	核电项目	巴基斯坦	巴基斯坦核电站项目

数据来源：中国"一带一路"网：https://www.yidaiyilu.gov.cn/info/iList.jsp?cat_id=10005；亚投行官网：https://www.aiib.org/en/index.html。

（五）进行"一带一路"气候投融资的主要机构

自"一带一路"倡议实施以来，中国的金融机构积极投身项目建设，为企业提供资金支持。据测算，"一带一路"地区基础设施投资缺口每年将超过6000亿美元[①]。如果在建设初期就考虑潜在气候风险，建设具有气候韧性的基础设施，其产生的环境效益将非常可观且可持续。参与"一带一路"建设的主要机构包括中资银行和中国倡议发起的国际金融机构。

1. 中资银行积极投入"一带一路"沿线气候投融资

（1）国家开发银行

国家开发银行在2017年完成"一带一路"重大国际规划22项，在沿线国家新增发放贷款176亿美元，融资支持沿线国家基础设施互联互通、产能和装备制造合作、金融合作和境外产业园区建设等。另外，国家开发银行实现人民币专项贷款授信承诺991亿元，发起设立中国–中东欧银联体，推动与上合银

---

① 详见：https://baijiahao.baidu.com/s?id=1597522887730681238&wfr=spider&for=pc。

三、"一带一路"沿线的气候投融资

联体、中国—东盟银联体、金砖国家银行合作机制等多双边金融合作①。

2017年国家开发银行成功发行首只5亿美元和10亿欧元中国准主权国际绿色债券，债券募集资金用于支持"一带一路"建设相关清洁交通、可再生能源和水资源保护等绿色产业项目，改善沿线国家（地区）生态环境，增强沿线国家（地区）应对气候变化能力。此外，国家开发银行通过私募形式在香港发行3.5亿美元"一带一路"专项债，创新内地与香港市场互联互通支持"一带一路"建设融资新模式。牵头主承销马来亚银行10亿元人民币"债券通"熊猫债，专项用于支持境内外"一带一路"项目建设，是东盟国家首只、中国债券市场首单"债券通"熊猫债②（见图3-4）。

图3-4 国家开发银行"一带一路"气候投融资情况

（2）中国进出口银行

2016年，中国进出口银行投资巴基斯坦卡洛特水电站项目正式开工。该项目规划装机容量72万千瓦，总投资金额约16.5亿美元，预计于2020年投

---

① 国家开发银行. 2017国家开发银行年度报告［R］. 2017.
② 国家开发银行. 2017国家开发银行可持续发展报告［R］. 2017.

入运营。项目建成后每年将为巴基斯坦提供约32亿千瓦时的清洁能源,将有效缓解巴基斯坦国内电力短缺的问题。

2017年12月22日,中国进出口银行2017年第一期"债券通"绿色金融债券在上海发行,期限3年,金额20亿元人民币,发行利率4.68%。多家银行包括中国银行新加坡分行及中国香港、欧洲等多家境外投资机构积极参与发行认购,参与认购金额5.2亿元,最终配售金额2.6亿元。该绿色债券募集资金将投向"一带一路"沿线国家清洁能源和环境改善项目,经独立第三方评估机构测算,募集资金投放将在二氧化碳、二氧化硫、氮氧化物减排方面获得良好的环境效益①(见图3-5)。

图3-5 中国进出口银行"一带一路"气候投融资情况

(3)中国银行

中国银行在"一带一路"沿线23个国家设有机构,是在"一带一路"沿线国家设立分支机构数量最多的中资银行。截至2017年末,中国银行共跟进"一带一路"重大项目逾500个。2015年至2017年,对"一带一路"沿线国家提供约1000亿美元的授信支持,并协助匈牙利政府成功发行首只募集资金明确用于"一带一路"的主权熊猫债②(见图3-6)。

---

① 中国进出口银行.进出口银行"债券通"绿色债上海成功发行,助力绿色"一带一路"全球互联建设[R].2017.

② 中国银行.中国银行2017年年度报告[R].2017.

三、"一带一路"沿线的气候投融资

图 3-6　中国银行"一带一路"气候投融资情况

（4）中国工商银行

中国工商银行卢森堡分行发行首只"一带一路"气候债券，发行额等值 21.5 亿美元，全球投资者超额认购。此债券分三笔发行，覆盖美元和欧元两个币种。债券募集资金将投向于可再生能源、低碳及低排放交通。截至 2017 年末，中国工商银行累计支持"一带一路"项目 358 个，合计承贷金额 945 亿美元，2017 年新增承贷项目 123 个，承贷金额 339 亿美元，在"一带一路"沿线 20 个国家和地区拥有 129 家分支机构①（见图 3-7）。

图 3-7　中国工商银行"一带一路"气候投融资情况

---

① 中国工商银行. 中国工商银行 2017 年年度报告［R］. 2017.

2. 中国倡议发起的金融机构承担"一带一路"沿线气候投融资重要角色

亚洲基础设施投资银行（AIIB，以下简称亚投行）与丝路基金是"一带一路"建设的主要资金来源。两者在投资方式上有所不同，亚投行着重于债权投资，主要通过发放贷款参与项目；而丝路基金运营模式偏向于股权直接融资，投资期限较长。两者还可以衍生出更多资金募集方式，为"一带一路"沿线国家募集更多项目资金①。同时，亚投行的资金来源主要是各国政府出资，是一种政府行为，而丝路基金则主要针对有资金且有投资意愿的主体，这意味着丝路基金可以吸收民间资本参与到"一带一路"建设中。

（1）亚投行

亚投行是首个由中国倡议设立的多边金融机构，重点支持基础设施建设。截至2018年12月，亚投行有87个成员国或候选成员国，其中已经认缴股本的区域内成员国有44个，区域外成员国有24个②。

亚投行是"一带一路"倡议建设的重要资金来源。截至2018年12月，亚投行官网列出的获批项目一共31个，涉及融资金额62.95亿美元，其中与气候相关的投资项目有19个，融资金额44.11亿美元，项目个数占总投资项目的60%以上，融资金额占总融资金额的70%以上。项目涉及可再生能源、绿色交通、城市废弃物处理、污水处理等领域③。

2018年亚投行在"一带一路"沿线国家新增8个投资项目。新增项目中除在印度的国家投资与基础设施资金项目通过投资FoF吸引机构投资者的民间资本，从而减少印度基础设施部门的股权融资缺口外，其余7个项目均投资于气候减缓与适应项目，其中清洁能源类项目3个，水资源类项目2个，清洁交通类项目2个。

从投资项目所在国看，新增项目中有3个项目投向印度，项目总投资为6.95亿美元；两个项目投向土耳其，项目总投资为8亿美元；分别在埃及、印度尼西亚、孟加拉国有一个项目投资，投资额分别为3亿美元、2.5亿美元和0.6亿美元（见图3-8）。

---

① 详见：http://opinion.hexun.com/2017-05-20/189268794.html。
② 详见：https://www.aiib.org/en/about-aiib/governance/members-of-bank/index.html。
③ 详见：https://www.aiib.org/en/projects/approved/index.html。

三、"一带一路"沿线的气候投融资

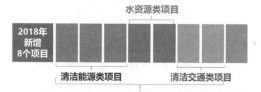

图 3-8 亚投行气候投融资项目情况

（2）丝路基金

丝路基金 2014 年 12 月 29 日成立于北京，重点围绕"一带一路"建设推进与相关国家和地区的基础设施、资源开发、产能合作和金融合作等项目。丝路基金以股权投资模式为主，债权、基金、贷款等多种投融资方式相结合，为"一带一路"建设提供投融资服务。

丝路基金的资金规模为 400 亿美元和 1000 亿元人民币，其中外汇储备（通过梧桐树投资平台有限责任公司）、中国投资有限责任公司（通过赛里斯投资有限责任公司）、中国进出口银行、国家开发银行（通过国开金融有限责任公司）的出资比例分别为 65%、15%、15% 和 5%[1]（见图 3-9）。

丝路基金自 2014 年成立以来，积极参与到"一带一路"沿线项目建设中，项目分布覆盖面广，类型多样，包括基础设施、资源开发、产能合作和金融合作等。目前丝路基金已在"一带一路"沿线跟踪并储备项目 100 多个，范围覆盖俄蒙中亚、东南亚、南亚、西亚北非、中东欧等重点区域和国家[2]。

从丝路基金的投资项目看，现有气候投资主要为清洁能源类项目，包括

---

[1] 详见：丝路基金官网，http://www.silkroadfund.com.cn/cnweb/19854/19858/index.html.

[2] 详见：http://www.financialnews.com.cn/zgjrj/201704/t20170428_116703.html.

水电、天然气、清洁燃煤发电、光伏发电。在投资方式上，丝路基金倾向采用股权投资的方式参与项目。截至2018年3月底，丝路基金超过70%的投资额为股权投资。

图3-9　丝路基金投资方出资比例

对一些资金体量大的项目，丝路基金在进行股份认购的同时也配套发放贷款，采取"股权+债权"投资的创新模式实现企业与丝路基金的双赢。一方面，股权投资降低项目的资产负债率，使项目更易获得资金支持；另一方面，债权投资使基金有相对保险和稳定的收益，降低了单独股权投资的风险。"股权+债权"的模式兼顾了投资的风险与收益，例如在巴基斯坦卡洛特水电站项目、俄罗斯亚马尔液化天然气一体化项目、迪拜哈翔清洁燃煤电站项目中，丝路基金均采取了"股权+债权"的投资模式。

丝路基金在项目建设过程中也注重对项目所在地环境的保护，将项目对当地生态系统和生物的负面影响降至最低，采用最先进和严格的生产技术标准，使项目满足当地碳排放标准，并在施工环节中采取相应保护措施。

案例2

丝路基金投资中东首座清洁燃煤电站

迪拜哈翔清洁燃煤电站位于阿拉伯联合酋长国迪拜以南30多公里的海边，这是中东地区首座清洁燃煤电站，也是"一带一路"框架下中东地区首个中资企业参与投资、建设和运营的电站项目。迪拜哈翔清洁燃煤电站项目是迪拜政府能源战略规划的重点项目，该电站由4台600兆瓦机组组成，总投资金额约33亿美元，首台机组计划在2020年3月投入商业运行。预计在2023年全面投入运营后，将满足迪拜总电力需求的20%。

丝路基金作为项目投资方不仅提供贷款，还推动中资金融机构参与银团贷款，共带动了约16亿美元中方银行信贷。丝路基金采取"股权+债权"的投资模式，一方面股权投资降低了项目的资产负债率，使项目更容易获得资金的支持；另一方面债权投资与股权投资形成了相互补充和配合，能够达到投资风险与收益的平衡。

在生产技术上，哈翔项目采用最先进的燃烧、脱硝、除尘和脱硫技术，保证电厂运行期间排放的粉尘、硫化物及氮氧化物指标优于世界同类型机组，减少大气污染物的排放，施工队伍通过采用降噪和温度控制工艺，使电厂远程受点噪声低于45分贝，循环排水点周围500米内温度提升不超过2℃，浇筑混凝土下面需要有保护层，减少对地下水的影响。项目采用的超临界清洁燃煤技术符合欧盟最严格的工业碳排放标准，有利于当地绿色环保和节能减排[①]。

此外，项目采取高环保要求，在施工过程中注重对当地生态系统的保护。迪拜哈翔项目码头施工区域位于当地自然保护区内，哈电国际积极履行企业在当地的社会责任及环保义务，对施工区域内的28000株珊瑚进行移植和培养，将潟湖区内鱼类全部转移至大海。在每年海龟繁殖期，按照与迪拜环保组织共同制定的方案，对海龟进行监控与保护。施工方专门设立了水质与空气质量监测站，用以实时监测施工区域的环境参数[②]。

[①] 详见：http://finance.jrj.com.cn/2017/05/09052022447390.shtml。
[②] 详见：http://big5.xinhuanet.com/gate/big5/big5.news.cn/gate/big5/silkroad.news.cn/2018/0727/105672.shtml。

四、绿色气候基金（GCF）的发展情况和融资方案分析

根据《联合国气候变化框架公约》确定的发达国家与发展中国家"共同但有区别"的责任原则，发达国家应承担减少温室气体排放的主要责任，并有义务向发展中国家提供新的、额外的资金支持。绿色气候基金（GCF）在这项原则下应运而生。

（一）GCF的发展情况

1. GCF成立的目的是为发展中国家应对气候变化筹集资金

绿色气候基金（GCF）是由《联合国气候变化框架公约》（UNFCCC）缔约方的194个国家成立的，作为公约财务机制的一部分，旨在为发展中国家减缓和适应气候变化提供资金。

设立GCF的提议最先提出于2009年哥本哈根气候变化大会。《哥本哈根协议》中规定，由发达国家出资建立绿色气候基金，用于支持发展中国家应对气候变化的行动。协议初步规定发达国家应在2010年至2012年出资300亿美元作为快速启动资金，到2020年每年提供1000亿美元的资金支持。随后GCF在2010年举行的坎昆气候变化大会上被确立。在2015年巴黎气候变化大会上，联合国重申了发达国家对GCF的出资义务，并提出将聘请技术专家对气候资金情况进行评估，督促发达国家完成每年1000亿美元的出资承诺，但也并未对其有强制性的出资任务分配。GCF的发展历程如图4-1所示。

四、绿色气候基金（GCF）的发展情况和融资方案分析

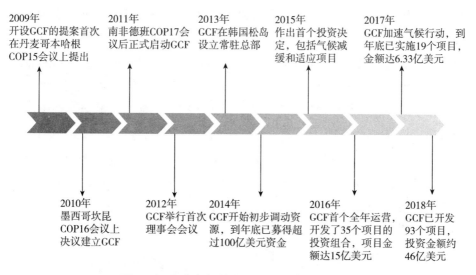

图 4-1 绿色气候基金（GCF）发展历程图

2. GCF 下设董事会、秘书处、受托管理人

GCF 下设董事会、秘书处、受托管理人等职能部门。董事会的主要职权是监督基金的运作，核准业务模式、利用模式和供资结构，核准具体业务政策和指南，任命秘书处执行主任，遴选和任命受托管理人等。

GCF 秘书处由董事会任命并对董事会负责，负责基金的日常运作，提供财务、行政、法律方面的支持。2012 年，韩国仁川被选为秘书处所在地，韩国承诺 2019 年前每年为 GCF 注资 100 万美元，并在 2014—2017 年以信托基金的形式向 GCF 援助 4000 万美元。

受托管理人须按照国际标准管理其金融资产，编写财务报表和其他相关报告。目前，世界银行为 GCF 的临时受托人，将在基金运行三年后接受审计。

3. GCF 的出资模式主要有赠款、优惠贷款、股权和担保四种

目前 GCF 从 43 个国家共筹集了 103 亿美金，其项目投资通过执行机构进行，已经授权了 75 个执行机构，包括世界银行、联合国开发计划署、联合国环境规划署、亚洲开发银行等国际金融机构和国际组织，也包括汇丰银行等私人部门机构，其中有两个是中国的机构，分别是中国清洁发展机制基金管理中心和对外经济合作办公室。

GCF 设立后一直重视对气候适应项目的投资，致力于使减缓和适应类项

目的投资比重均衡。截至2018年12月，已开发了93个项目，涉及投资金额46亿美元，累计可减少14亿吨二氧化碳排放，其中39%是减缓气候变化项目，25%是适应类项目，36%是两者都有涉及。在地理分布上，项目遍布全球各主要大陆，有40个项目位于亚太地区，36个项目位于非洲，20个项目位于拉丁美洲和加勒比地区，6个项目位于东欧。然而，GCF至今在中国还没有项目投资。

GCF对这些项目的出资模式有赠款、优惠贷款、股权和担保四种，投资金额分别为21亿美元、19亿美元、4.2亿美元和8000万美元，分别占总投资额的47%、42%、9%和2%（见图4-2、图4-3）。

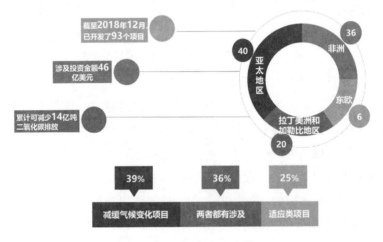

图4-2　GCF投资项目情况

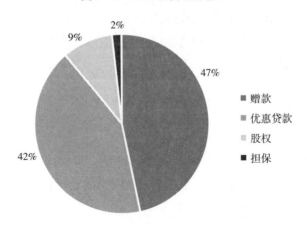

图4-3　GCF各出资模式占比

四、绿色气候基金（GCF）的发展情况和融资方案分析

4. GCF 的主要特征

（1）GCF 运作过程中注重平等原则

GCF 重点关注极易受气候变化影响的地区的需求，包括最不发达国家、小岛屿发展中国家和非洲国家。GCF 在运作过程中注重缔约国平等与地域平等。UNFCCC 公约下的其他基金，如全球环境基金（GEF）曾受到发展中国家的质疑，主要是由于在这些基金的运作机制下，边缘海岛国家以及最不发达国家几乎对资金投向没有话语权。GCF 在运作过程中充分考虑到发展中国家的参与情况，在过渡委员会筹建时纳入了 25 位发展中国家成员，其中来自最不发达国家和小岛屿发展中国家的成员各有 2 位，另有 15 位发达国家成员，给予发展中国家充分的话语权。GCF 正式成立以后，董事会由发展中国家与发达国家缔约方各 12 名成员组成，2 名联合主席由董事会成员从内部选出，分别来自发达国家与发展中国家，充分体现了平等原则。

（2）GCF 向政府公共部门和私人部门融资

相比于其他气候基金仅从公共部门筹资，GCF 专门设立了私营部门机制（PSF）。私营部门在应对气候变化方面可以发挥很大的作用，但私营资金逐利性强，较难保证其负责任地进行气候投资。GCF 采取谨慎的态度吸收私人资本，充分评估其对环境、社会和发展产生的影响，保证其对减缓和适应气候变化产生正面效应。在 GCF 已投资的 93 个项目中，40% 的资金来自私营部门。PSF 不仅能为 GCF 募得更多资金，也能带动发展中国家私营部门参与到减缓和适应气候变化活动中。

（3）发达国家出资比例悬而未决，GCF 正在探索其他融资途径

虽然发达国家承诺为 GCF 融资，但因其义务不具有强制性，且缺少具体细化的资金分摊要求，发达国家出资意愿低，拖延注资义务。快速启动资金官方网站的数据显示，发达国家承诺的 2010—2012 年出资 300 亿美元快速启动资金，目前只兑现了 36 亿美元。

历届联合国气候变化大会虽然强调设立 GCF 的重要性，也受到各缔约方的一致赞同，但具体到发达国家间的出资比例，依然没有统一的说法和口径。在 2018 年 12 月刚闭幕的 COP24 波兰卡托维兹气候变化大会上，全体缔约方通过了《巴黎协定》实施细则，提供了 2025 年后新的气候资金兑现方法，承诺到 2020 年发达国家每年向发展中国家提供 1000 亿美元的气候资金，德

国和挪威政府在大会上承诺对GCF增加一倍的气候资金投入,但缔约方依旧未达成具体的发达国家出资比例要求。

因此,除发达国家的资金援助承诺外,GCF目前还在积极动员有能力的发展中国家为其注资,并推动公私部门合作,吸收社会资本,进行项目开发。

(二) GCF融资责任的分摊机制分析

一直以来,对于GCF如何筹集资金没有具体方案,GCF正面临融资不足的困境。GCF属于国际资金募集,资金来源于发达国家,主要流向发展中国家。对于国际资金筹集,融资分摊存在某种共性,即由于缺乏一个超越各国主权的权力机构,融资完全依赖于各国间的谈判协商,而最终方案一般为各国妥协的结果。因此,有必要对现有的国际资金筹集模式进行总结,并探索其对GCF融资分摊的借鉴意义。

下面将讨论GCF融资任务如何在发达国家间进行分摊的问题,从历史责任和支付能力两个维度出发,借鉴联合国会费分摊(UN)、联合国官方发展援助计划(ODA)以及全球环境基金(GEF)等现存的国际资金筹集经验,讨论不同融资机制下GCF融资责任的分摊效果。考虑到不同国家会有不同的方案偏好,采用投票理论对不同方案进行加权,得到一种兼顾多种分摊思想的PSC融资责任分摊体系,其也将对新兴经济体参与气候融资情景进行分析,并评估美国退出《巴黎协定》对气候融资发展的影响。

1. 单指标GCF融资责任分摊方案

(1) 基于历史责任分摊方案(HR)

HR原则假定各国的GCF融资份额正比于其历史排放责任,选取各地区1850—2015年累积CO_2排放量为计算依据。结果发现,如果GCF筹资义务基于HR原则,则美国需要出资44.30%,欧盟需要出资25.57%,英国需要出资8.36%,日本需要出资6.82%。在HR原则下美国资金压力较大,需要承担近一半的筹资义务。

(2) 基于经济实力的分摊方案(AP)

AP原则假定各国的GCF融资份额与其经济能力挂钩,选取GDP作为经济能力的表征。为消除年份间波动,以2010—2015年均GDP为计算基准。结果发现与HR类似,如果GCF筹资任务基于AP原则分摊,美国是最大的出资方，其筹资份额为35.38%。紧随其后的是欧盟，该地区需要承担30.38%的融资任务。日本和英国的捐资份额分别为11.67%和6.37%。其他发达国家也需要承担一定筹资任务，但份额均比较小。

（3）基于联合国会费分摊经验（UN）

UN方案假设各成员国的出资份额正比于其联合国会费缴纳水平，以2013—2015年UN年均会费为计算依据。结果是：如果GCF筹资采用UN会费分摊经验，36.61%的融资任务需要欧盟承担，28.23%的筹资任务需要由美国承担，日本和英国的出资份额分别为13.90%和6.65%。与HR和AP方案不同，在UN方案下欧盟是最大的捐资方。

（4）基于官方开发援助融资分摊经验（ODA）

ODA是发达国家与发展中国家间开展的一项国际援助计划，资金来源于发达国家并流向发展中国家，这种资金流动特点与GCF一致。假设各出资成员分摊份额正比于其ODA援助金额，以2010—2014年均援助水平为计算依据，可得欧盟是最大的捐助方，其出资份额为39.82%。美国、英国和日本分别需贡献22.81%、11.40%和7.78%。

（5）基于全球环境基金融资分摊经验（GEF）

GEF方案假设各出资国GCF融资水平正比于其GEF出资水平，即为全球环境基金注资较多的国家也会愿意承担较多绿色气候基金筹资义务。在此方案下，43.47%的融资任务需要欧盟承担。美国和日本筹资份额分别为14.83%和16.49%。与前四种方案相比，此种方案下欧盟资金压力最大。

2. 兼顾多种分摊思想的PSC融资责任分摊方案

（1）PSC方案设计思想

上述五种方案都可为GCF融资提供宝贵借鉴，但捐资国需要通过谈判决定采取哪一种。而在实际谈判中，出于国家利益考量，不同国家会有不同的决策偏好。为了降低出资国在GCF融资谈判中出现的分歧，需要在不同方案间寻求一种平衡，并尽可能兼顾不同国家的利益诉求。

本研究基于投票理论，提出一种偏好得分妥协法（PSC）的多指标分摊体系。表4-1以HR、GEF和ODA为例说明PSC方法的加权思想。此时，一种备选方案若被某国所偏好，那么该方案就会得票，票数即为该国的人口水平。可知，三种方案在发达国家内部的偏好权重依次为47.09%、47.70%

和 5.21%。基于 PSC 方案,美国需要出资 30.34%,欧盟需要出资 35.03%,日本需要出资 11.85%,英国需要出资 8.51%。剩余部分中,加拿大和澳大利亚分别需要出资 4.36% 和 1.93%。

表 4-1　　　　　　　PSC 融资责任分摊方案设计思想

	备选方案			备选方案投票(百万人)			投票结果
	HR	GEF	ODA	HR	GEF	ODA	责任分摊
美国	44.33	28.23	16.76	0.00	0.00	315.47	30.34
日本	6.82	13.90	16.59	127.47	0.00	0.00	11.85
加拿大	3.46	3.83	5.30	34.94	0.00	0.00	4.36
澳大利亚	1.86	2.10	1.99	23.00	0.00	0.00	1.93
新西兰	0.20	0.32	0.21	4.46	0.00	0.00	0.21
瑞士	0.32	1.34	3.20	8.06	0.00	0.00	1.75
挪威	0.29	1.09	1.73	5.05	0.00	0.00	1.02
英国	8.37	6.65	8.86	0.00	63.93	0.00	8.51
欧盟	25.59	36.61	44.17	374.28	0.00	0.00	35.03
韩国	1.71	2.56	0.24	0.00	0.00	50.08	1.05
墨西哥	2.04	2.36	0.35	0.00	0.00	119.63	1.25
ODV	5.02	1.01	0.60	0.00	0.00	99.60	2.70
得票	—	—	—	577.26	63.93	584.78	100.00

注:(1) ODV 是其他发达国家,包括智利、巴拿马、冰岛、哥伦比亚、列支敦士登、秘鲁和摩纳哥。

(2) 假定备选方案为 HR、GEF 和 ODA。

(2) 基于 PSC 的 GCF 融资责任分摊方案集

基于 PSC 的加权思想,以 HR、AP、UN、ODA 和 GEF 为备选方案,并允许不同方案间自由组合,可以得到 31 种 GCF 融资分摊的潜在方案,如图 4-4 所示。可以发现,尽管不同方案差异性明显,但美国、欧盟、日本和英国始终是最为重要的出资方,共需贡献超过 80% 的筹资任务,这表明 GCF 对上述四个地区的依赖性较强。

四、绿色气候基金（GCF）的发展情况和融资方案分析

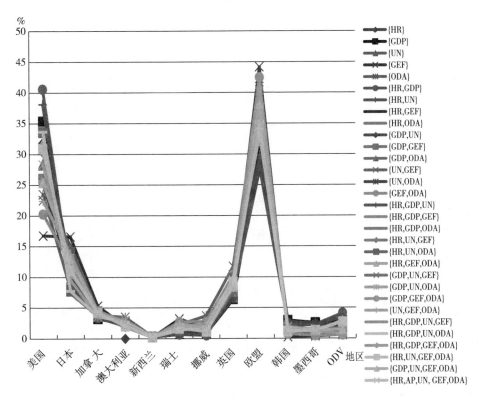

图4-4 基于PSC的GCF融资责任分摊方案集

3. GCF融资分摊方案的有效性评估

为评估31种方案的有效性，选取GCF第一次增资期（2015—2018年）为对比基准，引入相关关系和最短距离函数两种测度方式，结果如图4-5所示。可知两种测度方式下备选方案有效性排序完全相同，表明该结果具有较强的稳健性。在欧式距离情景下，与参考基准最为接近的方案为{HR，GEF，ODA}，两者间的距离为0.06，而在相关系数情景下，与参考基准最为相似的方案亦为{HR，GEF，ODA}，相关系数高达0.99，这表明两者间具有高度正相关关系。此外，研究发现最不相似或者最不贴近的是HR方案，这表明GCF融资分摊不应只考虑各国的历史排放责任，还要兼顾各方的经济能力水平。

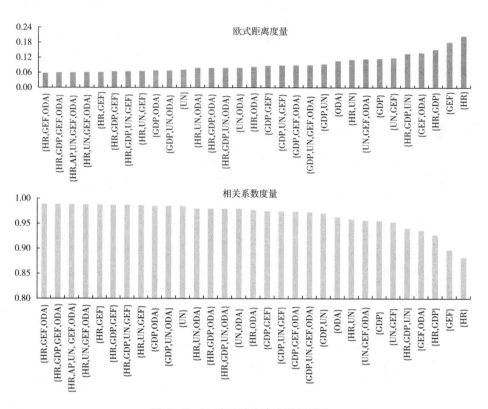

图4-5 31种设计方案有效性排序

4. 敏感性分析

（1）美国退出气候融资的影响

这部分主要讨论当美国退出气候融资后对其他捐资国产生的影响，以有效性最强的｛HR，GEF，ODA｝方案为例进行分析。如图4-6所示，美国退出气候融资将会增加其他国家的捐资负担，尤其对欧盟的影响最大。具体来看，欧盟需要承担48.83%的筹资任务，日本需要承担15.53%的筹资任务，英国需要承担12.97%的筹资任务。与美国不退出相比（表4-1），欧盟出资份额增加13.80个百分点，英国增加4.46个百分点，日本增加3.68个百分点。

四、绿色气候基金（GCF）的发展情况和融资方案分析

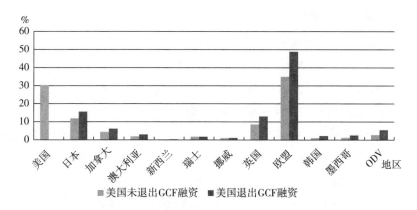

图 4-6　美国退出气候融资前后对其他捐资国的影响

（2）气候融资成员拓宽至金砖国家

这部分讨论当 GCF 捐资成员拓宽至金砖国家时的影响，同样以 {HR, GEF, ODA} 为例进行分析。如图 4-7 所示，即便 GCF 融资成员拓宽至金砖国家，发达国家仍然是绿色气候基金的主要捐资方，累计需要承担 94.43% 的出资任务，金砖国家需要承担 5.57% 的筹资义务。此时，中国需要贡献 2.54%，俄罗斯需要贡献 1.36%，巴西需要贡献 0.51%，印度和南非则分别需要贡献 0.85% 和 0.31%。这表明，{HR, GEF, ODA} 方案并不会给金砖国家带来明显的资金负担，进一步显示出该方案设计的合理性。

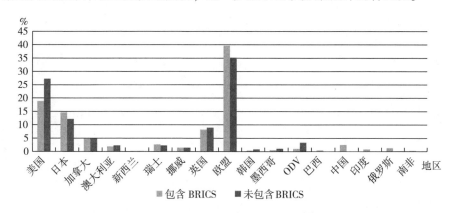

图 4-7　气候融资成员拓宽至金砖国家的影响

5. 分析结论

（1）不同融资机制的责任分摊差异较大，如美国在 GEF 方案下出资份额最小，日本和欧盟在 HR 原则下分摊份额最小。

（2）采用PSC方法对HR、UN和GEF复合可以得到一种有效性最强的方案，其分摊效果较为贴近现实，此时欧盟需要出资35.03%，美国需要出资30.34%，日本和英国分别需要出资11.85%和8.51%。

（3）若气候融资成员拓宽至金砖国家，该地区总共需承担5.57%的出资任务，表明GCF捐资对发达国家依赖性较强。

（4）美国退出《巴黎协定》会显著增加其他国家的融资负担，欧盟捐资份额将增加14个百分点。

因此，GCF融资责任分摊不仅需要考虑历史排放责任，也要兼顾各捐资国的经济能力，对两者进行权衡时需要考虑不同国家的方案偏好，这有助于提升方案的可接受性。本研究建议采用PSC方法对HR、GEF和ODA进行复合，构造得到GCF筹资义务的分摊方案，这不仅最为贴近GCF第一次增资的区域贡献，也不会给新兴经济体带来过重的财务负担，是对公平与效率的兼顾。

五、多边开发银行在气候融资中的角色

继 2002 年墨西哥蒙特雷和 2008 年卡塔尔多哈之后,联合国第三次发展筹资问题国际会议于 2015 年在埃塞俄比亚首都亚的斯亚贝巴举行。会议通过的《亚的斯亚贝巴行动议程》提出:"我们鼓励多边开发金融机构建立一个审视自己角色、规模和功能的机制,使其能够适应和充分响应可持续发展议程。①"

由于多边开发银行的使命、规模和影响力,他们在实现可持续发展目标方面发挥着重要作用。多边开发银行引导营利性金融市场走向政策目标,在公共资本和私人资本之间搭建了桥梁②。多边开发银行的特点决定了他们可以证实某些方法、行业和地区的可行性,而其他投资者可能不会这样做。为了实现气候金融领域的政策目标,多边开发银行既可以直接扩大气候投资,也可以将气候因素纳入其一般融资的考虑中。2017 年,G20 领导人在汉堡发布联合声明,呼吁多边开发银行加强气候融资行动③。此外,多边开发银行于 2017 年 12 月在巴黎召开的一次全球首脑会议上发表了一份联合声明,承诺其资金流动与《巴黎协定》2℃目标相一致④。

(一)多边开发银行在促进可持续发展方面具有优势

亚的斯亚贝巴会议汇集了众多利益相关方,讨论如何为可持续发展目标融资。与会者包括 50 多位国家元首、200 多名部长以及联合国和其他政府间

---

① Addis Ababa Action Agenda. *Addis Ababa Action Agenda*(AAAA)*of the Third International Conference on Financing for Development*[R]. Addis Ababa, Ethiopia:United Nations, 2015.
② MDBs. *From Billions to Trillions:MDB Contributions to Financing for Development*[R]. 2015.
③ G20. *G20 Leaders Statement "Shaping an interconnected world"*[R]. Hamburg, Germany:G20, 2017.
④ MDBs. *Joint Statement by the Multilateral Development Banks at Paris, COP*21[R]. Paris, France:MDBs, 2017.

机构，如世贸组织、知名企业、民间社会组织和其他利益相关方，会议提出了《亚的斯亚贝巴行动议程》（以下简称 AAAA）。

在促进可持续发展方面，AAAA 强调了多边开发银行的五个主要特点①：（1）长期性和稳定性；（2）反周期性；（3）优惠性；（4）专有技术和技术援助；（5）撬动私人资本。

表 5-1 总结了多边开发银行的特点、优势和典型实例。

表 5-1　　　　　　　多边开发银行的特点总结及其优势

多边开发银行的特点	优势	典型实例
长期性和稳定性	多边开发银行对项目作出长期担保，在项目与行业层面和地域层面都可以提供稳定的投资	多边开发银行的平均贷款期限为 20—30 年 世界银行和亚洲开发银行长期致力于通过清洁空气倡议来改善亚洲城市空气质量
反周期性	在其他投资者不愿投资的领域进行战略性投资	统计数据表明，在国际金融危机之后，多边开发银行扩大了气候融资的规模，弥补了许多国家在这一领域的国家预算削减②
优惠性	能够以各种形式提供优惠，包括向战略领域提供混合融资	亚洲开发银行的绿色金融优惠条款和美洲开发银行的绿色信贷额度提升，都旨在提高绿色项目的可融资性
专有技术和技术援助	无论是就具体项目还是就整个国家或地区而言，规模大、投资经验丰富的多边开发银行拥有比其他投资者更全面的专业知识	欧洲投资银行和世界银行采取措施，促进绿色债券的发行 国际货币基金组织和世界银行建立伙伴关系，进行税务诊断

---

① Addis Ababa Action Agenda. *Addis Ababa Action Agenda（AAAA）of the Third International Conference on Financing for Development*［R］. Addis Ababa, Ethiopia：United Nations, 2015.

② OECD. "CRS：*Aid activities*", *OECD International Development Statistics（database）*［R］. Paris, France：OECD, 2016.

续表

多边开发银行的特点	优势	典型实例
撬动私人资本	通过金融政策支持和金融机制,多边开发银行可以降低机构和项目层面的认知风险	多边开发银行每投资1美元就能撬动2—5美元的私人资本投资[①],而发达国家对发展中国家提供的公共气候资金这一指标仅为0.34美元[②]

(二) 气候融资发展面临多种挑战

从根本上说,气候融资缺口是由低投资回报率导致的,这种低投资回报率产生的原因可能是较低的收入,也可能是较高的成本。在收入方面,低碳项目通常可以获得与非低碳项目同等的收入,甚至可能获得针对性补贴,例如可再生能源发电的补贴。但是从成本方面来说,低碳项目对投资者的吸引力较小。这些成本通常源于不成熟的技术导致的技术风险、不稳定的政策支持导致的政策风险、缺少实践经验带来的高操作成本、项目和贷款时间期限不匹配以及金融市场的高交易成本。气候融资发展面临的挑战概述见表5-2。

表5-2　　　　　　　　气候融资发展面临的挑战概述

挑战类别	主要内容	现有应对措施
体制框架	A. 政治、经济和环境不稳定 B. 政策和监管的不确定性 C. 不合适的补贴和电价 D. 国有企业不公平的竞争环境 E. 进入的监管门槛较高	通过国际论坛和机制,对政策执行实施从战略到监管再到地方层面上的改革

---

① MDBs. *From Billions to Trillions*: *MDB Contributions to Financing for Development* [R]. 2015.
② OECD & CPI. *Climate Finance in 2013–2014 and the USD 100 billion goal* [R]. Paris, France: OECD, 2015.

续表

挑战类别	主要内容	现有应对措施
项目融资者	A. 项目开发成本高 B. 交易成本高 C. 专项资金提供者对低碳项目的竞争 D. 过分强调短期回报 E. 投资组合的限制	建立对可持续项目成本增加的补偿机制,提高投资者对环境、社会和治理的认识,将气候融资的要求纳入项目
项目开发者	A. 对气候融资机制的认识有限 B. 在利用非传统金融方面缺乏经验 C. 组织低碳项目的能力有限 D. 缺乏透明和全面的项目渠道 E. 缺少资金和可行的商业模式	改进公共或私营机构提供的指导和培训,并开发知识交流和筹资的平台
金融市场	A. 缺乏低碳资产类别 B. 缺乏专门的资金和融资工具 C. 风险错配 D. 非货币化的正环境外部性 E. 缺乏数据 F. 准确评估低碳项目风险的能力较低	通过平台、标准和第三方评估提高信息的可用性和质量,开发创新的金融工具,并建立协调气候融资流程和标准的国际对话机制

(三) 多边开发银行在气候融资中起重要作用

在气候融资领域,多边开发银行在 2016 年总共提供了 270 多亿美元,其中 77% 用于减缓,23% 用于适应气候变化①。为实现《联合国气候变化框架公约》的 2020 年承诺目标,多边开发银行在 2013 年至 2015 年大力提供气候融资支持,其支持额度占发达国家对发展中国家总支持额度的三分之一以上②。尽管多边开发银行的累计支持额度比较大,但是考虑气候融资在多边开发银行总融资中占的比例也是十分重要的,同时需要考虑多边开发银行在非气候投资组合中的气候份额。

图 5-1 显示了一些多边开发银行的气候融资比例及其 2020 年目标。根

---

① MDB Joint Report. *Joint Report on Multilateral Development Banks Climate Finance* [R]. 2016.
② OECD & CPI. *Climate Finance in 2013 – 2014 and the USD 100 billion goal* [R]. Paris, France: OECD, 2015.

据承诺，多边开发银行将提供发达国家对发展中国家气候融资支持总额的40%①。这意味着大多数多边开发银行需要在短时间内取得重大进展，才能实现他们的目标，因此他们必须在融资方向上进行快速转型。除了积极的气候融资目标外，多边开发银行还使用负面清单来排除项目。例如，2017年世界银行行长金墉在巴黎峰会上宣布，世界银行将不再为油气资源的开发提供资金，从2010年开始世界银行就已经停止了燃煤相关项目。

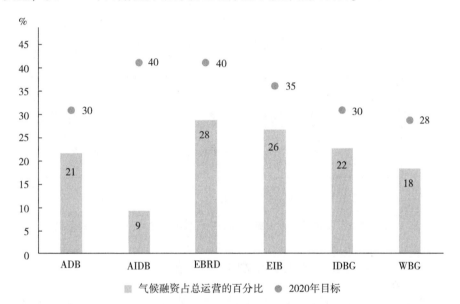

图5-1 多边开发银行2016年气候融资比例和2020年目标②

考察多边开发银行投资组合中的气候份额也相当重要。为评估与2℃目标的相符性，世界资源研究所分析了世界银行（WB）、国际金融公司（IFC）和亚洲开发银行（ADB）的气候融资行动（这三家机构的融资额占多边开发银行总融资额的三分之一），得出的结论是：17%的融资是与2℃目标相一致的，57%的融资在特定条件下一致，22%的融资是有争议的，3%的融资与目

---

① OECD. Investing in Climate. Investing in Growth [R]. Paris, France：OECD, 2017.
② MDB Climate Finance, The Good, the Bad and the Urgent. Washington DC, USA：WRI [R]. 2017.

标不一致①。虽然这个研究存在一定的局限性，但是表明了多边开发银行需要改变现有政策，以确保其气候政策目标。经济合作与发展组织（OECD）的统计进一步表明了这一点，如图5－2所示，虽然气候融资在基础设施融资的能源类别中所占比例较高，但在其他基础设施类别中所占比例仍然很小。在2013年到2015年，多边开发银行基础设施融资总额中有三分之一是用来缓解和适应气候变化的。

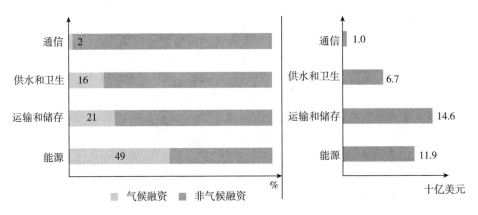

图5－2 多边开发银行基础设施各类别中气候融资额与总融资额的比较②

（四）应采用多种方式继续推进气候融资

多边开发银行越来越重视气候金融，这一点在多边开发银行的战略性文件中表现明显，如长期计划、对气候相关领域投资占比的承诺。另外，绿色债券发行量迅猛增长也是这一点的具体体现。虽然早期的多边开发银行基础授权文件侧重于发展、增长和转型，但他们允许通过当前战略文件来增加绿色业务数量。而那些新成立的多边开发银行通常将绿色业务作为其初期业务的核心支柱。

多边开发银行可以考虑采取以下措施，以进一步发挥其在气候金融领域的作用。

---

① WRI. Financing the Energy Transition：Are World Bank, IFC, And ADB Energy Supply Investments Supporting A Low – Carbon Future? [R]. Washington DC, USA：WRI, 2017.

② OECD. Investing in Climate. Investing in Growth [R]. 2017.

1. 重视撬动私人资本

多边开发银行可以在撬动私人资本、为环境可持续发展提供资金方面发挥重要作用。未来可持续发展所需资金大部分都将由私营部门提供,而多边开发银行拥有史上最高的资金杠杆能力。根据多边开发银行的《从数十亿到数万亿》报告,多边开发银行的杠杠率为2—5美元①,远高于南北气候融资等其他融资渠道②。虽然一些多边开发银行已经在动员私人资本参与,但需要更多的多边开发银行参与进来,并将其作为所有业务的基础。

2. 实施内部碳定价

由于不是所有国家都有外部碳定价,多边开发银行可以通过内部碳定价来将外部因素内部化,从而减轻项目融资中的实际和转型风险。内部碳定价为确定每个项目的碳足迹提供基础,而定价方法的不同会直接影响到项目的可融资性。为了与《巴黎协定》目标保持一致,碳定价领导联盟建议到2020年碳价范围为40—80美元/吨③。内部碳定价不仅可以为项目评估提供直接的成本激励,也可以与改进的风险评估方法结合使用,以全面识别成本和风险方面的气候因素。

3. 给予有针对性而非广泛的优惠支持

虽然多边开发银行的优惠融资业务可以鼓励对应某些政策目标的融资,但这种支持可能会产生不利的市场扭曲效应。建议与开发性金融机构的指导文件一致,应在气候金融领域实施精准的优惠支持,而非提供广泛的补贴④。气候金融中的许多案例都表明,相比长期的优惠融资,项目投资方更需要短期的优惠利息或优惠期限以及多边开发银行的技术支持。

4. 加强环境风险评估

尽管多边开发银行具有丰富的金融经验,但外部环境的迅速变化仍对目前的风险评估方法提出了挑战。除了传统风险,如项目的内部金融特性和外

---

① MDBs. *From Billions to Trillions: MDB Contributions to Financing for Development* [R]. 2015.

② OECD & CPI. *Climate Finance in 2013 – 2014 and the USD 100 billion goal* [R]. Paris, France: OECD, 2015.

③ Carbon Pricing Leadership Coalition. *Report of the High – Level Commission on Carbon Prices* [R]. 2017.

④ Development Finance Institutions. *DFI Working Group on Blended Concessional Finance for Private Sector Projects* [R]. 2017.

部风险因素外,多边开发银行必须采用新的方法来评估新的风险,如2017年G20绿色金融研究小组和金融稳定委员会气候相关工作组强调的气候风险以及过渡性环境相关风险。此类评估方法应充分包括资产等级数据、影响度量、潜在情景、管理影响以及其他特定情景变量。

5. 深化规模经济下的多边开发银行合作

多边开发银行可以通过相互间合并的融资解决方案来提高气候融资的有效性。虽然多边开发银行在政策建议、定价和融资方式上的竞争是良性的,但这也有可能导致不完善的金融体系发展①。如果多边开发银行的不同融资解决方案存在很多重叠,那么就可以通过规模经济来提高效率,最终实现《釜山有效发展合作伙伴关系》下的伙伴关系原则②。

6. 增强非气候融资方案向气候的转化

由于多边开发银行日益强调为气候变化领域提供资金,在某些情况下,可以对偏重其他优先事项的现有的、成功的融资方案进行修改,使其包含气候融资。《亚的斯亚贝巴行动议程》鼓励在2015年后的发展议程中更多地进行政策和实践的更新,每个多边开发银行都可以分析各自的融资解决方案,以确定如何扩大气候融资。

---

① Brookings. The New Global Agenda and the Future of the Multilateral Development Banking System [M]. Washington DC, USD: Brooking Institution, 2018.

② OECD. The Busan Partnership for Effective Development Co-operation [R/OL]. 2011. Retrieved from: http://www.oecd.org/development/effectiveness/busanpartnership.htm.

六、政策建议

(一) 在全球气候治理格局发生改变的新形势下继续发挥"引领者"作用

《巴黎协定》之后全球气候治理格局已发生改变,从欧盟倡导的"自上而下"的温室气体减排量强制性分配转变为中美所倡导的"自下而上"的国家自主贡献(INDC)模式。INDC体现了各缔约国愿意承担的减排承诺和具体行动,是相对宽松灵活的温室气体减排模式,更适合目前的全球应对气候变化形势,有利于气候谈判的顺利进行。

目前,欧盟的气候治理领导力逐渐减弱,美国特朗普政府宣布退出《巴黎协定》,而中国已经从气候治理体系的"跟随者"转变为"引领者"。在哥本哈根会议期间,中国成为国际气候谈判的重要参与者,代表发展中国家发声。之后,以中国为代表的"基础四国"迅速崛起,对气候谈判的走势和治理模式的转变起到了关键作用,并坚持了"共同但有区别的责任"这一基本原则,将其落实到《巴黎协定》中。《巴黎协定》时期,中国不仅在自主贡献文件中提出了实际的减排目标,体现了大国责任,还协调其他大国共同推进《巴黎协定》的签署。2016年9月,中国利用G20峰会的主场优势促成中美正式加入《巴黎协定》,极大地推进了协定的生效和实施。2018年12月,卡托维兹气候变化大会完成了《巴黎协定》大部分内容实施细则的谈判,取得了巨大的成果,中国为大会取得成功作出了重要贡献,成为国际气候谈判达成的关键推动力量。

在《巴黎协定》大部分内容实施细则已经达成,而美国因为政府更迭导致气候政策转向的大背景下,中国坚持在气候治理方面的大国责任,因势利导形成更强的国际领导力是明智的战略选择。这有利于主导国际规则制定和提高对国际事务的话语权,提高应对国际事务的能力和借鉴国际经验解决内部事务,有利于促进与其他国家的合作和进一步增加中国对发展中国家的影响力,也符合国内产业结构升级和环境治理的需求。

中国在应对气候变化中发挥引领作用应坚持以下三个方面原则。

第一,灵活务实、多方共赢。在应对气候变化方面,采取积极自信的气候外交政策,以开放的姿态与发达国家、发展中国家开展合作,同时更加灵活务实,更多地从全球视角出发,寻求中国在国际舞台上应对和引领全球气候变化问题所应发挥的作用,最大化中国及其他国家利益,寻求多方共赢。

第二,顺势而为、树立形象。中美是最大的碳排放国,承担了较大的减排压力。目前,作为发达国家的美国因为政府更迭而导致气候政策转向,成为全球应对气候变化进程的消极力量。当前美国是众矢之的,中国只需要继续坚持自己近年来的气候外交政策原则,因势利导,不需要额外的实际投入即可达成发挥更强的国际领导力的效果,这是中国主导应对气候变化、积极参与和影响国际事务、树立自身形象的机会。值得一提的是,应避免为体现大国责任而盲目地提供资金,要坚持"共同而有区别的责任",一方面在国际博弈中最大化中国利益,另一方面适当体现中国的责任担当。

第三,战略合作、重视双边。"基础四国"合作和"南南合作"是中国构建领导力的主要支撑。增强"基础四国"合作有利于增强气候信息共享,协调谈判立场,并深化在新能源减排技术等方面的务实合作。"南南合作"可以同中国的"一带一路"倡议全面结合,有利于输出中国在新能源方面的产能优势,并结合亚投行的投资布局,开展低碳投资。在多边合作框架下,建议高度重视与"基础四国""一带一路"国家的双边合作,在有利于减排、保护环境的基础设施投资、新能源投资等绿色金融领域展开深度合作。

(二) 积极利用国际气候资金推动国内减缓和适应气候变化

1. 站在发展中国家立场推动对"1000亿美元"目标的实现

"1000亿美元"气候资金应该是在UNFCCC框架下,发达国家向发展中国家提供的新的、额外的资金,并应以公共资金为主。中国应联合其他发展中国家,坚持这一原则,推动国际共识的尽快落地,使"1000亿美元"目标有衡量的基础。

另外,中国需坚持自身的发展中国家立场,坚持中国的资金受助方立场。

国际气候资金与发展中国家国内的自主减排资金投入不应该混淆，不能因为中国在自主减排方面的突出成果就剥夺中国的资金受助权。目前中国的气候风险仍然较高，减缓和适应领域的资金缺口巨大，毫无疑问需要国际资金和技术的支持。

2. 增强自身的能力建设，加强接收方数据的监测和统计

目前对气候资金的追踪报告均出自发达国家或者联合国，而作为气候资金的接收方，还未有发展中国家发布追踪、监测气候资金流的官方报告。虽然发展中国家严重质疑发达国家公布的气候资金数据的公信力，但由于没有自身统计数据，无法对气候资金的接收数据进行去伪存真。中国应深入参与到国际气候资金核查、报告和监测（MRV）体系建设之中，并联合其他发展中国家，从资金接收国角度，建立气候资金MRV体系。这样既有利于改进国际数据统计体系，又能为发展中国家在国际气候谈判中提供有力依据。有了自身的统计数据，才能更好地挑战和质疑有问题的数据。另外，建立和完善气候资金MRV体系，可以帮助中国明确资金缺口，了解资金协调与管理的差距，有利于在国家层面制定更加符合国情需求的气候资金战略和政策。

3. 重视GCF的示范效应，学习国际气候治理的丰富经验

目前GCF还没有在中国投资的项目，中国应积极争取GCF的资金支持。在GCF的75个执行机构中，已经有2个中国的机构具备了通过国内机构申请资金支持的条件。虽然GCF所能提供的资金相对有限，但其示范效应十分突出，尤其是对于落后地区的能力建设方面，可以通过项目执行，引入国际上丰富的知识和经验，加强国际资本与国内资本的合作。另外，还可以通过GCF的示范项目，分享和传播其气候治理的先进经验，给国内其他气候治理的项目提供优秀的范例，提高项目所在地应对气候变化的能力。

GCF常用的投资形式有赠款、优惠贷款、股权和担保，其中赠款和优惠贷款占总投资额的比重超过90%。中国可考虑以这些形式与GCF展开合作，并探索债券等新型投资方式，吸引国际资金投资中国气候融资领域。鉴于GCF只对内在创收好的项目进行资金支持，中国可以考虑除申请GCF资金外，同时申请潜在的其他公共或私人资金，以扩展GCF申请项目方案的融资优惠条件水平。

4. 积极影响以 GCF 为代表的国际气候资金机制运营规则和技术规则的制定

如何平衡分配资金是 GCF 目前面临的最大挑战。平衡分配体现在资金均衡地分配在气候适应和减缓领域，以及公平地在发展中国家进行分配。

中国要积极影响以 GCF 为代表的国际气候资金机制的规则制定，避免发达国家以及代表其利益的国际组织片面地制定国际规则。气候资金中相关技术规则是气候资金性质在资金实体运行中的重要体现，在坚持 GCF 等气候资金独立性、使用效果与持续长期资助挂钩的基础上，技术标准成为实际影响发展中国家能够获得何种气候资金及其数量的重要因素。中国应积极影响规则制定，在国际场合理性发声，并坚持此类技术标准要体现发展中国家的普遍意志，保证特定资金均衡分配。

（三）大力推动市场化碳交易并发展碳金融

1. 推动碳现货市场健康稳定发展

碳现货市场的发展是碳期货以及其他碳金融产品创新的基石，其健康稳定发展取决于以下几方面条件的完善。

一是完善的市场顶层设计。应尽快完成碳市场建设的立法程序，并完善相关配套制度，建立科学、统一、执行性强的方法学，做好基础设施建设，提高数据的准确性和信息的透明度，充分发挥 cap－and－trade 机制的作用，真正发挥市场定价的功能，实现减排目标。

二是稳定积极的交易政策。首先政策要稳定和明确，例如配额发放机制、CCER 政策等；其次要鼓励具有一定规模的交易，这样才能形成合理碳价；此外要适当引入多元化的市场参与主体，合理的市场规模以及充分的流动性，是碳市场能够有效发挥定价功能的基础，但为了防止个人投资者盲目入市，加剧市场波动，在市场不成熟时要谨慎引入个人投资者。

三是灵活的价格调控机制。建议建立碳市场的价格稳定调控机制，可以建立政府储备池和碳市场平准基金，合理设定碳价调控区间，并依托碳配额储备与平准基金建立公开市场操作机制；还可以设置价格涨跌停及最大持仓量限制制度，限制当日最高涨跌的百分比和配额持有者的最大持仓量。

2. 推动碳期货市场的建设，引入金融监管

期货市场是现货市场的有效补充，有风险管理、套期保值、促进价格发

现、提供流动性和套利机会等作用。碳交易期货市场能够更加有效地配置碳资产，因此，研究启动碳期货市场建设，扩大碳市场的层次体系，实现碳现货市场与期货市场互利互动，能够有效推动碳排放权交易体系的良好建设。

同时，应当加强金融监管，宏观层面和微观层面均要做到审慎监管，施行综合防范措施，确保风险的可防范性和减少风险的放大性。由于证监会对于市场机制、价格调控、未来衍生品的引入、风险防控等一系列问题有着丰富的经验，建议在监管中引入证监会，进行联合监管。

3. 加强利益相关方的碳金融能力建设

利益相关方主要包括政府、企业和金融机构。碳市场是人为规定而非自发生成的市场，受政策影响较大，因而政府作为政策制定者，配备一定的碳金融知识储备，才能在制定政策时掌握好"度"，既保证碳市场的稳定运行，避免出现过多投机现象，又能增加碳市场的活跃度。企业和金融机构作为碳市场的主要参与者，只有提高其碳金融能力，才能有力催生碳金融衍生品的发展和促进流转，从而刺激整个市场健康良性活跃的发展。

就市场需求来说，碳金融方面的专业人才缺口很大。因此需要设计全面的碳金融课程，引进权威的师资力量，对政府、企业、金融机构等利益相关方进行定期培训，使他们熟悉业务的流程和规则，培养在职人员的理论知识水平和技术水平。另外，还要加强宣传和科普，改变传统观念，增强社会整体对节能减排和碳金融的认识。

4. 鼓励碳金融产品创新试点

为推动碳金融体系的层次性发展，在稳步前行的基础上，政府可出台激励措施，适当鼓励碳金融产品创新试点。金融机构应丰富产品类型，鼓励创新形式，探索贴合企业和机构实际需求，具备可操作性、简便性、流通能力较强的碳金融产品。对于风险较高的碳金融衍生品，近期可以推动场外产品，如远期、掉期产品，逐步推动场内金融产品，如期货、期权产品的推出，并注意有效监管，防范风险。

（四）加强"一带一路"沿线的气候投融资

1. 继续推进有利于应对气候变化的基础设施建设

"一带一路"沿线国家普遍存在基础设施建设落后和社会经济发展水平

不高的问题。因此投资有利于应对气候变化的基础设施建设，不但能为这些国家经济发展提供基础，增添力量，解决其燃眉之急，而且也有利于沿线生态环境的保护。应加大对沿线基础设施项目建设的生态环保服务与支持，推广铁路、城市轨道交通、城乡公路运输等清洁交通和清洁能源、绿色建筑项目，在项目建设中推动水、大气、土壤和生物多样性等领域的环境保护，提升绿色低碳化建设水平。

2. 加大清洁能源方面投资

"一带一路"沿线各国蕴含体量巨大的可再生能源，包括风能、太阳能、水能和潮汐能等，开发空间巨大。比如中亚、西亚多国日照时间长，光照强度大，具有丰富的太阳能开发潜力，另外这些国家地域辽阔，风能充沛，十分有利于风能发电；而东南亚、南亚地区多国拥有大量瀑布、河流等水资源，地势差明显，有利于开发水电。"一带一路"沿线国家可加大对清洁能源的投资，因地制宜地发挥沿线国家巨大的能源优势，在环境友好的同时，也可解决沿线国家能源供应不足的问题。

3. 推动投融资模式创新

现阶段中国在这些国家的投资项目，普遍为清洁交通类基础设施建设和清洁能源类项目。这些项目建设周期较长，投资回报周期也较长，如果采用传统的股权投资很难在短时间内获得投资收益。因此探索新的融资模式，在保证投资风险可控的前提下获得最大化长期和短期收益结合的投资组合，是吸引更多投资者投入"一带一路"建设的有效措施。比如丝路基金创新的"股权＋债权"融资模式，在进行项目股份认购的同时也配套发放贷款，兼顾投资的风险与收益，长期收益和短期回报。

4. 构建"一带一路"气候投融资统计体系

现阶段中国对"一带一路"地区已经进行了许多气候投融资项目建设，这些项目既利于增强当地应对气候变化能力，又能解决当地能源短缺、交通运输系统不发达等民生问题，应该大力提倡。中国应构建"一带一路"气候投融资项目统计体系，对此类项目进行专项统计和管理，建设中国自己的气候投融资专项数据库，为气候变化国际谈判提供数据支撑，也利于更有针对性地进行"一带一路"气候投融资。另外，可对项目库中具有显著气候效应的项目进行宣传和推广，形成示范效应。

5. 加强气候合作平台建设

中国应推动气候合作平台建设，提供项目支撑服务。目前已有多个国际合作平台如上海合作组织、中非合作论坛、澜沧江—湄公河合作机制等，可充分发挥现有双边、多边环保国际合作机制，构建气候合作网络，方便各国分享气候治理经验。另外，应加快创新气候合作模式，建设政府、智库、企业、社会组织和公众参与的多元合作平台，发挥各方优势，完善国际气候治理体系。

2018 China Climate Financing Report

INTERNATIONAL INSTITUTE OF GREEN FINANCE, CUFE

RESEARCH CENTER FOR CLIMATE AND ENERGY FINANCE, CUFE

About the Report

International Institute of Green Finance, CUFE (IIGF)

International Institute of Green Finance (IIGF) of Central University of Finance and Economics (CUFE), is known as the first international research institute in China whose goal is to promote the development of green finance. IIGF, grew out of the Research Center for Climate and Energy Finance, is one of the executive members of Green Finance Committee (GFC) of China Society of Finance and Banking while it has built an academic relationship with the Ministry of Finance. The IIGF aims to cultivate the economic environment and social atmosphere with the spirit of green finance and to build the domestic first-class, the world's leading financial think tank with Chinese Characteristics.

Research Center for Climate and Energy Finance, CUFE (RCCEF)

Founded in September 2011, RCCEF has issued China Climate Financing Report foreight consecutive years. Based on generalized concept of global climate financing, RCCEF has established an analytical framework of climate financing flow and created a model of China's climate-related financing demands. Besides, RCCEF does in-depth analysis on international climate capital governance as well as China's climate financing development on a yearly basis and has accumulated a series of research findings in this area. Based on long-term trust and cooperation, RCCEF co-established an academic partnership with the Ministry of Finance.

About the Report

Instructor

WANG Yao	Professor, Director General, PhD Tutor, International Institute of Green Finance, CUFE
	Deputy Secretary General, Green Finance Committee of China Society for Finance and Banking

Authors

CUI Ying	Senior Researcher, Research Center for Climate and Energy Finance, CUFE
HONG Ruichen	Researcher, Research Center for Climate and Energy Finance, CUFE
Mathias Lund Larsen	Researcher, Research Center for Climate and Energy Finance, CUFE
CUI Lianbiao	Invited researcher, International Institute of Green Finance, CUFE
ZHANG Shiyu	Research Assistant, Research Center for Climate and Energy Finance, CUFE
TIAN Xiaoye	Research Assistant, Research Center for Climate and Energy Finance, CUFE
QIAN Qingjing	Research Assistant, Research Center for Climate and Energy Finance, CUFE

Foreword

2018 is a significant year in global response to climate change following the adoption of the Paris Agreement in 2015. The United Nations Intergovernmental Panel on Climate Change (IPCC) released Special Report: Global Warming of 1.5℃, assessed the effects of global warming at 1.5℃ and 2℃, and proposed possible reduction paths. The 1.5℃ report describes a better world in which the temperature rise is controlled within 1.5℃, while the UNEP's Emissions Gap Report 2018 warns that 2017 Global greenhouse gas emissions reached a new high of 49.2 billion tons of carbon dioxide equivalent. From the perspective of emission reduction technology, the probability of achieving the 1.5℃ temperature control target is becoming lower, countries have to work harder than the current emission reduction work to achieve the goal of 2℃ temperature control.

In 2018, there were some negative events against climate change. The outbreak of the "yellow vest" campaign in Paris, France to protest against the government's imposition of a fuel tax, Swiss Parliament voted against the amendments to the Carbon Dioxide Act aimed at achieving the Paris Agreement goals, and the German Coal Exit Committee recommended to postpone the closure of the first batch of coal-fired power plants from 2020 to 2022. More and more evidence show that some EU countries are likely to miss their 2020 and 2030 emission reduction targets.

The 24th United Nations Climate Change Conference in Katowice, held under these negative impacts, is a revitalization of the global response to climate change. The participants finally agreed on most of the implementation rules of the Paris Agreement, reiterated the obligation of developed countries to contribute to climate finance.

Foreword

China has greatly promoted the negotiation in Katowice Conference. The Chinese delegation mediated among representatives of all parties, effectively promoted the negotiation process and contributed Chinese experience to global climate governance. Under the background of the withdrawal of the United States from the Paris Agreement, China has increasingly played the leading role in international climate governance and take the responsibility of a big country.

China's Belt and Road Initiative launched in recent years advocates low-carbon development and green investment. Countries along the route have fragile ecological environment and poor ability to cope with climate change. Climate financing for these countries can not only meet the needs of local people, improve their life quality, but also effectively eliminate high-energy consumption and high pollution capacity with green production capacity, prevent further deterioration of the environment and reduce climate risks.

At the same time, China's domestic climate financing is also developing steadily. At the end of 2017, China announced the launch of the national carbon market construction, which is now in the period of infrastructure construction. Once the national carbon market is completed, it will become the largest carbon market in the world, with significant emission reduction effect. The vigorous development of the green financial system represented by green bonds and green credit will also provide climate finance and demonstration effect for China's climate financing.

The deadline for global action on climate change after 2020 agreed by the Paris Agreement is very close. However, there is still no clear division of tasks for all countries to deal with climate change. Governments need to spare no efforts to promote the progress of international negotiations and strengthen national actions. In view of China's leading experience in the field of green finance and the improved international influence, China will play a pivotal role in Global Climate Governance, affecting the overall development of climate finance.

List of abbreviations

Abbreviation	Full name	Chinese translation
A		
ADB	Asian Development Bank	亚洲开发银行
AfDB	African Development Bank	非洲开发银行
AIIB	Asian Infrastructure Investment Bank	亚洲基础设施投资银行
B		
BOC	Bank of China	中国银行
BFIs	Bilateral Financial Institutions	双边金融机构
C		
CCER	China Certified Emission Reduction	中国核证自愿减排量
CDB	China Development Bank	国家开发银行
CDM	Clean Development Mechanism	清洁发展机制
CDMF	Clean Development Mechanism Fund	清洁发展机制基金
CER	Certification Emission Reduction	核证减排量
CIFs	Climate Investment Funds	气候投资基金
COP	Conferences of the Parties	联合国气候变化框架公约缔约方大会
CPI	Climate Policy Initiative	气候政策倡议
D		
DFIs	Developmental Finance Institutions	发展性金融机构
E		
EBRD	European Bank for Reconstruction and Development	欧洲复兴开发银行
EC	European Commission	欧盟委员会
EIB	European Investment Bank	欧洲投资银行
ERPA	Emission Reduction Purchase Agreement	减排量购买协议
EU-ETS	European Union Emission Trading System	欧盟碳排放交易体系

Continued

Abbreviation	Full name	Chinese translation
EXIMB	the Export-Import Bank of China	中国进出口银行
F		
FAO	Food and Agriculture Organization	联合国粮食和农业组织
G		
GCA	Global Commission on Adaptation	全球适应委员会
GCF	Green Climate Fund	绿色气候基金
GDP	Gross Domestic Product	国内生产总值
GEF	Global Environmental Facility	全球环境基金
GGGI	Global Green Growth Institute	全球绿色发展署
I		
IADB	Inter-American Development Bank	泛美开发银行
IBRD	International Bank for Reconstruction and Development	国际复兴开发银行
ICBC	Industrial and Commercial Bank of China	中国工商银行
IEA	International Energy Agency	国际能源署
IFC	International Finance Corporation	国际金融公司
INDC	Intended Nationally Determined Contributions	国家自主贡献目标
IPCC	Intergovernmental Panel on Climate Change	联合国政府间气候变化委员会
IRENA	International Renewable Energy Agency	国际可再生能源机构
J		
JI	Joint Implementation	联合履约机制
L		
LDCs	Least Developed Countries	最不发达国家
M		
MDBs	Multilateral Development Banks	多边开发银行
MFIs	Multilateral Financial Institutions	多边金融机构
MIGA	Multinational Investment Guarantee Agency	多边投资担保机构
MRV	Monitoring, Reporting and Verification	国际气候资金核查、报告和监测
N		
NDBs	New Development Bank	新开发银行

Continued

Abbreviation	Full name	Chinese translation
O		
ODA	Official Development Assistance	官方发展援助组织
OECD	Organization for Economic Co-operation and Development	经济合作与发展组织
P		
PSF	Private Sector Facility	私营部门机制
PPP	Public-Private Partnership	政府与社会资本合作模式
R		
REDD+	Reducing Emissions from Deforestation and Degradation	减少毁林和森林退化造成的温室气体排放
S		
SPV	Special Purpose Vehicle	特殊目的实体
U		
UNDP	United Nations Development Programme	联合国开发计划署
UNEP	United Nations Environment Programme	联合国环境规划署
UNFCCC	United Nations Framework Convention on Climate Change	联合国气候变化框架公约
UN-REDD	United Nations Reducing Emissions from Deforestation and Forest Degradation	减少发展中国家毁林和森林退化所致排放量联合国合作方案
W		
WB	World Bank	世界银行
WBG	World Bank Group	世界银行集团
WMO	World Meteorological Organization	世界气象组织
WRI	the World Resource Institution	世界资源研究所

CONTENTS

I. Progress in global climate financing ······ 91
    (Ⅰ) Global climate finance has risen, but there is still a huge shortfall ······ 92
    (Ⅱ) Developed countries' climate finance commitment to developing countries is yet to be implemented ······ 97
    (Ⅲ) Climate financing for renewable energy is on the rise, but it still cannot meet the temperature control requirements ······ 100
    (Ⅳ) Blockchain technology starts to be applied in the field of climate financing ······ 104

Ⅱ. Progress in China's climate financing ······ 106
    (Ⅰ) National carbon market is announced to set up and local carbon pilots continue to operate ······ 106
    (Ⅱ) The driving force of green finance for climate financing is constantly emerging ······ 114
    (Ⅲ) PPP mode is steadily advancing in the field of ecological protection ······ 122
    (Ⅳ) China actively participates in international exchanges and cooperation, contributing to global control of climate change ······ 130

Ⅲ. Climate investment and financing along the Belt and Road ······ 137
    (Ⅰ) Background information of the Belt and Road Initiative ······ 138
    (Ⅱ) Climate characteristics of major countries along the Belt and

Road ·· 140
　（Ⅲ）The foundation to promote green investment along the Belt and
　　　Road ·· 145
　（Ⅳ）China's climate investment and financing projects for countries
　　　along Belt and Road Initiative ································ 150
　（Ⅴ）The main institutions for the Belt and Road climate investment
　　　and financing ·· 155

**Ⅳ. Development of the Green Climate Fund(GCF) and Analysis of
　Financing Options** ·· 166
　（Ⅰ）Development of the Green Climate Fund(GCF) ················ 166
　（Ⅱ）Exploring the Schemes for Green Climate Fund Financing ········ 171

Ⅴ. The Role of Multilateral Development Banks in Climate Finance ··· 181
　（Ⅰ）MDBs has advantages in promoting sustainable development
　　　·· 182
　（Ⅱ）Challenges within climate finance ································ 183
　（Ⅲ）MDBs are playing important roles in climate finance ············ 185
　（Ⅳ）MDBs should continue to promote Climate Finance in various
　　　ways ·· 187

Ⅵ. Policy Recommendations ································ 191
　（Ⅰ）Continue to play the role of "leader" under the new global
　　　climate governance ·· 191
　（Ⅱ）Actively use international climate funds to mitigate and adapt
　　　to climate change ·· 194
　（Ⅲ）Vigorously promoting domestic carbon market and developing
　　　carbon finance ·· 197
　（Ⅳ）Strengthening climate investment and financing along the Belt
　　　and Road ·· 199

I. Progress in global climate financing

Since the adoption of the Paris Agreement in December 2015, there has been no global agreement on the implementation details of the Paris Agreement. In December 2018, the 24th Conference of the Parties to the United Nations Climate Change was held in Katowice, Poland. After two weeks of negotiations, the participating countries finally agreed on the action rules for addressing climate change and formulated most of the contents of the Paris Agreement. The implementation rules laid the foundation for furthering the objectives of the Paris Agreement. The meeting reiterated that after 2020, developed countries will mobilize at least USD 100 billion of financial support to developing countries every year.

Before the conference, the United Nations Intergovernmental Panel on Climate Change (IPCC) issued the IPCC Global Warming 1.5℃ Special Report, pointing out that the current global temperature has increased by 1℃ compared with the pre-industrial level, the 1.5℃ rise of global temperature is likely to reach at 2030. The different climate impact between global temperature rise of 1.5℃ and 2℃ is significant. When the temperature rise reaches 2℃, it will have more destructive consequences such as habitat loss, ice cap melting, sea level rise and so on, which will threaten human survival and development and will also bring more significant damage to the world economy. To control the temperature rise below 1.5℃, the global climate action needs to be accelerated. The United Nations Environment Programme (UNEP) also issued the Emission Gap Report 2018, stating that there is still a large gap between the current situation and the goal of global warming not exceeding 2℃ set by the Paris Agreement. According to the current trend, the temperature control objectives of the Paris Agreement will inevitably fail, and even if countries implement their own Intended National Determined Contribution targets (INDCs) as planned, the average temperature of the Earth may rise more than 3℃ at the end of this century. Both reports indicate that climate negotiations must be intensified to encourage countries to make more stringent mitigation actions.

Climate financing is an important way to promote global emissions reduction. At present, the global climate capital flow has increased, but the funding gap is still huge. The USD 100 billion climate funding commitment of developed countries to developing countries has yet to be implemented. Renewable energy climate financing is growing, but it still cannot meet the temperature control requirements. In addition, the emerging of blockchain technology has begun to be applied in the field of climate finance.

(Ⅰ) Global climate finance has risen, but there is still a huge shortfall

At the end of November 2018, UNFCCC releasedthe 2018 Biennial Assessment and Overview of Climate Finance Flow before the Katowice Climate Change Conference in Poland, this report provided information on the climate fund flows through various channels in 2015 and 2016.

According to the UNFCCC, the quality and integrity of climate finance data has improved since 2016, and all agencies have improved methods for assessing climate funding data. The overall increase in global climate capital flows is significant. On a comparable basis, climate finance flows increased by 17% in the period 2015-2016 compared with the period 2013-2014. High-bound climate finance estimates increased from USD 584 billion in 2014 to USD 680 billion in 2015 and to USD 681 billion in 2016. The growth seen in 2015 was largely driven by high levels of new private investment in renewable energy, which is the largest segment of the global total. In 2016, a decrease in renewable energy investment occurred, which was driven by both the continued decline in renewable technology costs and the lower generation capacity of new projects financed. However, the decrease

Ⅰ. Progress in global climate financing

in renewable energy investment in 2016 was offset by an 8% increase in investment in energy efficiency technologies across the building, industry and transport sectors.

1. Flows from Annex II Parties to non-Annex I Parties[①]

Climate-specific finance report in biennial reports submitted by Annex II Parties shows climate fund has increased in terms of both volume and rate of growth since the previous biennal assessment. Whereas the total finance reported increased by just 5% from 2013 to 2014, it increased by 24% from 2014 to 2015(to USD 33 billion), and subsequently by 14% from 2015 to 2016(to USD 38 billion). Out of these total amounts, USD 30 billion in 2015 and USD 34 billion in 2016 were reported as climate-specific finance channeled through bilateral, regional and other channels; the remainder flowed through multilateral channels. From 2014 to 2016, both mitigation and adaptation finance grew in more or less equal proportions, namely by 41% and 45%, respectively.

---

① The United Nations Framework Convention on Climate Change divides Parties into three categories, namely Annex I Parties, Annex II Parties and developing country Parties. Annex I consists of 24 OECD countries, members of the European Community and 11 countries that have transitioned to a market economy. Annex II consists of 24 OECD countries and members of the European Community. Specifically, Annex II countries include Australia, Italy, Austria, Japan, Belgium, Luxembourg, Canada, the Netherlands, Denmark, New Zealand, European Community, Norway, Finland, Portugal, France, Spain, Germany, Switzerland, Greece, Turkey, Sweden, Iceland, the United Kingdom of Northern Ireland, Ireland and the United States of America. In addition to these, Annex I countries include Estonia, Belarus, Bulgaria, Czechoslovakia, Hungary, Romania, the Russian Federation, Latvia, Lithuania, Ukraine and Poland.

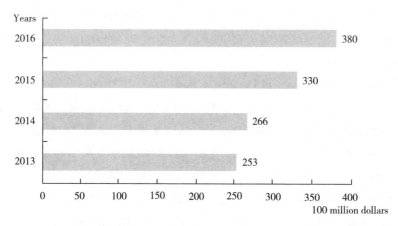

Figure 1 –1 Climate Funds from Annex II Countries to Non-Annex I Countries in 2013-2016

2. Multilateral climate funds

Total amounts channeled through UNFCCC funds and multilateral climate funds in 2015 and 2016 were USD 1.4 billion and USD 2.4 billion, respectively. The significant increase from 2015 to 2016 was a result of the Green Climate Fund(GCF) ramping up operations. On the whole, this represents a decrease of approximately 13% compared with the 2013-2014 biennium and can be accounted for by a reduction in the commitments made by the Climate Investment Funds, in line with changes in the climate finance landscape as the GCF only started to scale up operations in 2016.

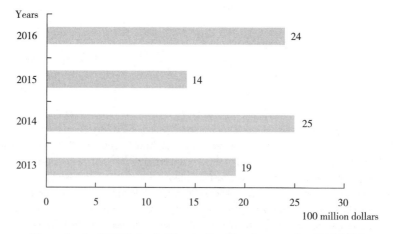

Figure 1 –2 Multilateral Climate Funding Amount for 2013-2016

3. Climate finance from multilateral development banks

MDBs provided USD 23.4 billion and USD 25.5 billion in climate finance from their own resources to eligible recipient countries in 2015 and 2016, respectively. On average, this represents a 3.4% increase from the 2013-2014 period.

The attribution of MDB finance flows to members of OECD-DAC, minus the Republic of Korea, is calculated at up to USD 17.4 billion in 2015 and USD 19.7 billion in 2016 to recipients eligible for OECD-DAC official development assistance.

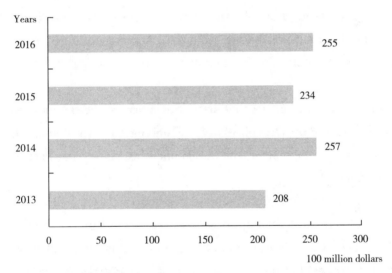

Figure 1-3 The amount of climate funds provided by the multilateral development banks from their own resources in 2013-2016

4. Private climate finance

The most significant source of uncertainty relates to the geographic attribution of private finance data. Although efforts have been made by MDBs and OECD since the 2016 BA to estimate private climate finance mobilized through multilateral andbilateral institutions, data on private finance sources and destinations remain lacking.

MDBs reported private finance mobilization in 2015 was USD 10.9 billion and increased by 43% the following year to USD 15.7 billion. OECD estimated USD 21.7 billion in climate-related private finance mobilized during the period 2012-2015 by bilateral and multilateral institutions, which included USD 14 billion from multilateral providers and USD 7.7 billion from bilateral finance institutions.

5. Domestic climate finance

Domestic climate expenditures by national and subnational governments are a potentially growing source of global climate finance, particularly as, in some cases, NDC submissions are translated into specific investment plans and domestic efforts to monitor and track the domestic climate expenditures are stepped up. However, comprehensive data on domestic climate expenditure are not readily available, as these data are not collected regularly or with a consistent methodology over time within or across countries.

6. Flows among countries that are not members of the Development Assistance Committee of the Organization for Economic Co-operation and Development, recipients eligible for official development assistance and Parties not included in Annex I

Information on climate finance flows among non-Annex I Parties is not systematically tracked, relying on voluntary reporting by countries through the OECD-DAC Creditor Reporting System and DFIs through IDFC that are based in countries that are not members of the Organization for Economic Co-operation and Development (non-OECD). Total estimates of such flows amounted to USD 12.2-13.9 billion in 2015 and USD 11.3-13.7 billion in 2016. This represents an increase of approximately 33% on average from the 2013-2014 period. New multilateral institutions include the Asian Infrastructure Investment Bank (AIIB) and the New Development Bank (NDB). Together, they provided USD 911 million to renewable energy projects in 2016. Although the climate capital flow has generally increased in the 2015-2016 biennial period compared to the previous biennial period, this level of funding is far from enough to meet the minimum investment required to mitigate and adapt to climate change. The International Energy

I. Progress in global climate financing

Agency's (IEA) projections show that in order to achieve the temperature control objectives of the Paris Agreement, the energy sector will need to invest USD 16.5 trillion in 2015-2030. The Global Green Development Agency (GGGI) estimates a climate financing gap of USD 2.5 to USD 4.8 trillion between 2016 and 2030. According to the United Nations Intergovernmental Panel on Climate Change (IPCC), the amount of adaptation funding for developing countries between 2010 and 2050 is between USD 70 million and USD 100 billion per year. The United Nations Environment Programme (UNEP) estimates that the annual demand for adaptation funds will be between USD 140 billion and USD 300 billion per year by 2030. By 2050, the annual demand for adaptation funds will increase to between USD 280 billion and USD 500 billion. Despite there is still large differences in estimates of climate investments required for private and public resources, the statistics of different agencies convey the same message: the amount of global climate funding needed is much larger than what is currently invested.

(II) Developed countries' climate finance commitment to developing countries is yet to be implemented

At the United Nations Climate Change Conference in Copenhagen in 2009, developed countries promised to provide developing countries with USD 100 billion each year to support their climate change actions by 2020 due to historical emissions. However, due to the fact that there has been no specific funding plan for this USD 100 billion, the implementation of funds is not satisfactory. Developed countries and developing countries have different opinions on whether the fund is "provided" or "mobilized", "public" or "public and private", whether it is "new and extra" outside the traditional ODA.

At the subsequent UN Climate Change Conferences held in each year, USD 100 billion in climate funding support has been a focus of climate negotiations. In December 2015, nearly 200 UNFCCC Parties adopted the new global climate agreement, the Paris Agreement, at the Paris Climate Conference. According to the agreement, developed countries will mobilize at least USD 100 billion in annual support to developing countries after 2020, new amounts will be determined by 2025 and will continue to increase. The Paris Agreement entered into force on 4th November 2016, and the USD 100 billion became the benchmark for developed countries to provide financial support to developing countries.

In December 2018, the 24th UN climate change conference in Katowice, Poland, launched negotiations and consultations on the implementation details of the Paris agreement, laying a foundation for the implementation of the Paris agreement after 2020. The Katowice meeting put in place the implementation details for most of the provisions of the Paris agreement. In terms of climate funding, it reiterated the financial support of developed countries to the lower limit of USD 100 billion US for developing countries.

The trend of climate capital flows shows that, theoverall flow of climate funds to beneficiary countries is increasing. International public climate funds flow through three channels to non-Annex I countries, namely multilateral climate funds, such as climate investment funds, green climate funds, etc., bilateral agencies, such as the French Development Agency, Germany's KFW bank, etc., multilateral development banks, for example, World Bank, European Investment Bank, Inter-American Development Bank, etc. The characteristics of the international public climate capital flow during 2015-2016 are shown in Table 1 −1.

I. Progress in global climate financing

Table 1-1 Characteristics of international public climate funding flows during 2015-2016

	Annual average financing (100 million USD)	Supported areas				Supported areas		
		Adaptation (%)	Mitigation (%)	REDD+① (%)	Cross-Domain (%)	Grant (%)	Concessional loan (%)	Other (%)
Multilateral climate funds	19	25	53	5	17	51	44	4
Bilateral climate funding	317	29	50	–	21	47	52	<1
MDB Climate Fund	224	21	79	–	–	9	74	17

Data Source: UNFCCC. 2018 Biennial Assessment and Overview of Climate Finance Flow[R]. 2018.

From the above summary of the international public climate funds, it can be seen that financial support for the mitigation area is still greater than support for other areas. For the adaptation field, bilateral climate funds are the largest (29%) for adaptation, and the proportions of multilateral climate funds and multilateral development banks for adaptation are 25% and 21% respectively. Grants and concessional loans are the main tools for providing climate funding. In the period 2015-2016, grants and concessional loans accounted for 95% and 99% of bilateral climate funds and multilateral climate funds. 74% of the funds through the multilateral development banks are issued in the form of concessional loans, and another 9% are grants.

There is no unified conclusion on the statistics of the performance of developed countries. The UNFCCC biennial report only lists the climate fund flow data of various channels, and does not give statistics on how to correspond to the USD100 billion financial commitment. However, as can be seen from the statistical information of the UNFCCC biannual report, there is still a big gap between the public financial support from developed coun-

---

① REDD+ refers to "Reducing greenhouse gas Emissions from Deforestation and forest Degradation in developing countries." The meaning of " + " is to increase carbon sinks.

tries to developing countries and the USD 100 billion target. Even if private sector funding is included, it still cannot reach USD 100 billion.

(Ⅲ) Climate financing for renewable energy is on the rise, but it still cannot meet the temperature control requirements

The global economy is shifting from a traditional fossil energy-based growth model to a model based on sustainable use of resources and energy, so the depth and breadth of research on renewable energyis extending in recent years. According to the calculation of IPCC, to fulfill global warming control within 1.5℃, the renewable energy power generation in 2050 should account for 70% to 85% of all electricity, all coal power plants need to be closed, and natural gas power generation also needs to use carbon dioxide recovery and storage technology.

In June 2018, the 21st Century Renewable Energy Policy Network (REN21) released *the 2018 Global Status Report on Renewable Energy*, summarizing the global development of renewable energy in 2017. The report shows that renewable energy installed capacity accounted for 70% of the global net installed capacity increase in 2017, an increase of 7% over 2016. The use of renewable energy in the heating, cooling and transportation sectors, which account for 80% of the global terminal energy demand, lags far behind the power industry and has insufficient transformational power. In the global context, renewable energy investments have the following four trends:

1. Developing countries dominate in renewable energy investment

A great majority of new investment in renewable energy comes from developing countries. Investment in renewable energy has been led by developing and emerging economies since 2015. China's investment in renewable energy is huge, rising 30.7% year-on-year in 2017. Led by China, developing countries invested USD 177 billion in 2017, accounting for 63% of global renewable energy investment. Meanwhile, the total investment from developed countries was USD 103 billion, a decrease of 19% from 2016. In terms of unit gross domestic product, the Marshall Islands, Rwanda, Solomon Islands, Guinea-Bissau and some other developing countries invest far more in renewable energy than in developed and emerging economies.

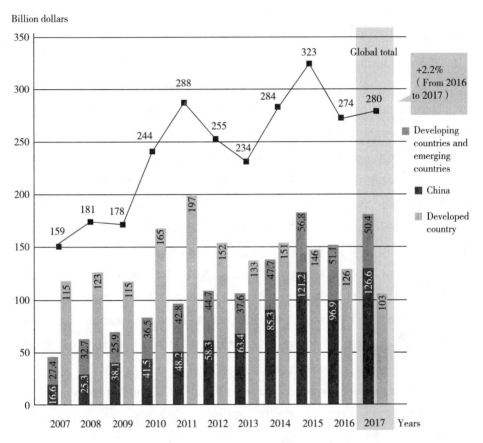

Data Source: The 21st Century Renewable Energy Policy Network (REN21). 2018 Global Status Report on Renewable Energy[R]. 2018.

Figure 1 – 4 New global investment in renewable energy and fuels from developed, emerging and developing countries in 2007-2017

2. Photovoltaic and wind power accounted for the majority of new installations

In 2017, about 178GW of new renewable energy capacity was installed worldwide, including about 98GW of photovoltaic power generation, an increase of 33% over 2016 and accounting for 55% of the total new capacity in 2017. New photovoltaic installations are larger than the net additions of fossil fuels and nuclear power combined, which achieve the highest in history. In addition, the global installed capacity of wind power reached 52GW, accounting for 29% of the total. Photovoltaic and wind power accounted for 84% of new renewable energy installations.

3. The transformation of renewable energy has been slower than expected

According to statistics, for every 3.7% increase in GDP, energy demand increases by 2.1%. [1] Although the world increasingly attaches importance to energy transformation and countries have also increased investment in energy transformation, the speed of renewable energy transformation in the power industry is far less than expected. Renewables lag far behind the power sector in heating, cooling and transport, which account for 80% of global end-use energy demand. In 2017, renewable energy accounted for only 10.3% of global heating energy consumption. Electricity provides only 1.3% of the energy needed for transport, and only about a quarter of that is from renewable sources. In addition, biofuels provide 2.8% of transportation energy needs, while 92% of the energy needs are still met by oil. The current pace of energy transformation is much slower than expected.

4. Renewable energy policies need to be adjusted and improved

Governments are adapting their renewable energy policies to the new economicsituation. In 2017, the governments of Switzerland, Denmark and Vietnam set new renewable energy targets. However, current renewable energy policies do not cover enough range. While 146 countries have renewable energy targets for the power sector, only 48 have renewable energy targets for heating and cooling, and 42 have renewable energy targets for transportation. If we want to achieve the goal of global warming control within 2 ℃ of the pre-industrial level in the Paris Agreement, and strive to target

---

[1] Renewable Energy Policy Network (REN21). The 2018 Global Status Report on Renewable Energy[R]. 2018.

1.5 ℃, the refrigeration, heating, and transportation sectors need to follow the transformation model of the power industry and transform energy at a faster rate.

To achieve energy transformation and fulfill climate and sustainable development commitments, governments need to play a guiding role in policy. Detail actions including establishing the correct policy framework, speeding up the transformation of the heating, cooling, and transportation sectors, promoting innovation in the field of backward and the development of new technology of renewable energy and reducing or ending subsidies on fossil fuels, should be done.

Energy demand continues to increase, with the growth of global economic and population. Energy demand and energy-related carbon dioxide emissions grew by 2.1% and 1.4% in 2017.[①] In order to solve the problems of increasing energy demand and energy transformation, the international community has been increasingly calling for energy reform. The international energy agency (IEA) estimates that USD 1 trillion per year is needed to fund the transition by 2050. Although climate finance flows have become more frequent over the past decade and are on the rise, there are still about 60% shortage below what is actually needed.[②] International renewable energy agency (IRENA) estimated that in order to reduce greenhouse gas emissions and limit global average temperature increase to 2 ℃ below, new low-carbon investment model must be unlocked, and benefit double growth must be achieved by 2030. Deployment of low-carbon solutions requires doubling current investment in renewable energy to USD 500 billion per year by 2020 and USD 900 billion per year by 2030.

[①] The International Energy Agency (IEA). Global Energy & CO_2 Status Report 2017 [R/OL]. 2017. https://www.iea.org/publications/freepublications/publication/GECO2017.pdf.

[②] Barbara K. Buchner, Padraig Oliver, Xueying Wang, Cameron Carswell, Chavi Meattle, and Federico Mazza. Global Landscape of Climate Finance 2017 [R]. [2017.10].

(Ⅳ) Blockchain technology starts to be applied in the field of climate financing

Blockchain refers to the use of a decentralized secure database to support key steps in a transaction, namely authentication, signing and payment. The identities of participants are stored on the blockchain so that the identity of the organization is algorithmic verified at each transaction, minimizing the risk of fraud. Blockchain creates "standard contracts" for specific transactions, which are translated into "smart contracts" by encapsulating their key terms in the blockchain distributed ledger and automatically executing and settling the contracts as long as the terms of the contracts are met. Using the automatic hosting feature inherent in the blockchain design, the parties will be paid in accordance with the terms of the contract when each phase of the contract is executed, eliminating the need for a trusted third party to manage payments. As a decentralized data storage technology, the data generated by blockchain cannot be tampered with or falsified, and any node contains all data information, so the data is extremely safe. Blockchain technology is now being used to simplify and accelerate the flow of transactions in climate finance, as it improves the security of transactions and opens the way to automation.

One of the barriers to climate capital flows is that many transactions are relatively small. Due to the decentralization of sources, the cost of third-party intermediaries is also paid, the transaction costs are high, and due to the opacity and asymmetry of information, donors and investors do not have sufficient confidence in the climate financing system, thus affecting the financing.

Blockchain can record transactions of any value, make the transfer of funds or assets between people or organizations transparent, and increase investors' interest in climate finance investment. At the same time, using blockchain can not only aggregate smaller transactions to form a large-scale transaction, but also combine investors to process larger transactions with faster transaction speed and lower cost by using algorithm aggregation and the execution of blockchain. Blockchain technology also makes it easy to match subcontractors to projects, grouping them by geography, technology and business operations. In addition, blockchain technology can also be used to run multiple procurement processes, including demand-side management(i. e., load reduction at high system loads), project refinancing, environmental bonds, and emission reduction purchase agreements (ERPAs).

The use of blockchain in climate finance is already happening on a small scale. CleanTek Market is developing a private, complete blockchain to support climate finance transactions. Gainforest is using smart contracts to incentivize small-scale amazon farmers to protect the rainforest. Smart contracts are automatically paid to farmers when remote sensing satellites verify that a particular forest has been successfully protected. Blockchain technology makes these transactions more transparent and can be trusted, and administrative costs are significantly reduced because there is no "middleman" to move money.

The application of blockchain technology in the field of climate finance will face the same problems as that in other fields. It is still in the initial stage and still needs to be improved and better matched with the application scenarios.

II. Progress in China's climate financing

In December 2018, at the 24th Conference of the Parties to the United Nations Framework Convention on Climate Change in Katowice, Poland, China hosted a forum about "China Climate Investment and Financing" at "China Pavilion". It introduces China's efforts in climate finance. In recent years, China has continuously exerted the leverage and support role of investment and financing for climate action, and actively carried out climate investment and financing activities. By promoting financial channels such as carbon emission trading markets, climate bonds, climate insurance, climate funds and so on, new funds have been injected into climate finance to incite social capital to climate change area.

(Ⅰ) National carbon market is announced to set up and local carbon pilots continue to operate

1. Pilot market construction progressed steadily

In October 2011, the National Development and Reform Commission approved the pilot projects of carbon emission trading in seven provinces and cities including Beijing, Tianjin, Shanghai, Chongqing, Hubei, Guangdong and Shenzhen. The seven pilots started trading from 2013 to 2014, and the market construction progressed steadily. In December 2016, two new pilots Sichuan and Fujian were newly introduced. The nine pilot operations provided experience for the nation's carbon market.

The policy design for each pilot has the same mechanism but reflects its own characteristics according to local condition. The relevant policies are compared as shown in the following table:

Table 2 – 1　　　　Comparison of carbon trading pilot policies

Pilot carbon market	Shenzhen	Shanghai	Beijing	Guangdong	Tianjin	Hubei	Chongqing	Sichuan*	Fujian
Trading platform	Shenzhen Carbon Emissions Exchange	Shanghai Environmental Energy Exchange	Beijing Environmental Exchange	Guangzhou Carbon Emissions Exchange	Tianjin Emissions Exchange	Hubei Carbon Emissions Trading Center	Chongqing Carbon Emissions Trading Center	Sichuan United Environmental Exchange	Straits Equity Exchange Center
Established time	2013.06.18	2013.11.26	2013.11.28	2013.12.19	2013.12.26	2014.04.12	2014.06.19	2016.12.16	2016.12.22
Allowance free distribution	Yearly allocation	Three-year distribution in 2013; Yearly allocation after 2016	Yearly allocation	Yearly allocation	Yearly allocation	Yearly allocation	Yearly allocation	Yearly allocation	Yearly allocation
Paid allowance distribution	Auction or fixed price for sale	Auction	Auction	Auction	Auction or fixed price for sale	Auction, the proportion does not exceed 30% of the government reserved allowance	—	Auction	Introduce at the right time

continued

Pilot carbon market	Shenzhen	Shanghai	Beijing	Guangdong	Tianjin	Hubei	Chongqing	Sichuan*	Fujian
Government reserved allowance	2% of total annual quota	—	5% of total annual quota	5% of total annual quota	—	10% of total annual quota	—	—	10% of total annual quota
CCER offset ratio and related regulations	No more than 10% of total allowance, must from Shenzhen's projects	No more than 1%, project should be built after January 1, 2013, Non-hydropower projects	No more than 5%, at least 50% of CCER should from Beijing	No more than 10%, at least 70% of CCER should from Guangdong	No more than 10%	No more than 10%, all projects should from Hubei	No more than 8%, all projects should from Chongqing	—	No more than 5%, all projects should from Fujian and non-hydropower
Other offset mechanism	—	—	Forestry carbon sink	PHCER	—	—	—	—	Fujian Forestry Carbon Sequestration (FFCER)
Individual investor	Yes	No	Yes	Yes	Yes	Yes	Yes	Yes	No

continued

Pilot carbon market	Shenzhen	Shanghai	Beijing	Guangdong	Tianjin	Hubei	Chongqing	Sichuan*	Fujian
Foreign investor	Yes	No	No	No	No	No	No	No	No
Up and down limit	10% (30% for block trades)	30%	Public transaction 20%	10% (listed bidding and listing)	10%	10%	20%	10% (30% for block trades)	Click to list 10%; contract transfer 30%
Financial derivatives	No	Yes	No	No	No	Yes	No	No	No

* Currently there is no trading in Sichuan market, thus the requirements on controlled enterprises, the number of government reseved allowances and the policies on CCER offset have not been announced.

Except for Sichuan Province, all pilots have started emission trading. The carbon market in Sichuan Province is still under construction in the early stage, it is expected that the simulation trading will be launched in 2019 and the emission trading will be launched in 2020.

The allowance allocation of the pilot carbon market is mainly based on free distribution, supplemented by paid distribution, and the proportion of paid distribution gradually increases. The most common way to paid distribution is by auction. In the allowance allocation process, the pilot governments will reserve a certain percentage of allowances for market stability purpose. In addition, each pilot sets up offset mechanism to offset part of allowances, but the pilot regulations are different for the setting of offset ratios and offsetting requirements. The setting of the offset mechanism has expanded the scope of the carbon market, enabling companies that have not joined the carbon market to join the energy conservation and emission reduction activities and increase market activity.

At present, all pilot carbon markets are mainly based on spot transactions. Some pilots have tried carbon finance and financing tools innovation, but the number of products is small and the amount is not large.

2. The national carbon market is in the period of infrastructure construction

On 19th December 2017, China announced the launch of the national carbon emission trading system, which caught global attention. On 18th December 2017, the National Development and Reform Commission issued the National Carbon Emissions Trading Market Construction Plan(Power Generation Industry), which clarified the guiding ideology and main principles of China's carbon market construction. This has important guiding significance for China to build a national carbon emission trading market.

The construction of national carbon market will go through three major stages: First is the infrastructure construction period. It will last for about one year, and complete the national unified data reporting system, registration system, trading system construction and carbon market management system. The second is simulation running period. It will also last about one

II. Progress in China's climate financing

year, to carry out the power generation industry simulation trading, to comprehensively test the effectiveness and reliability of the various elements of the market mechanism, to strengthen the market risk warning and prevention mechanism, and to improve the carbon market management system and support system. The third is deepening period when the national carbon market begins. During this period, carbon spot trading will be conducted among the power industry. After the stable operation of the power industry, it will expand industry coverage and trading varieties, and incorporate national voluntary emission reductions(CCERs) into the carbon market as soon as possible. The construction phase is shown below:

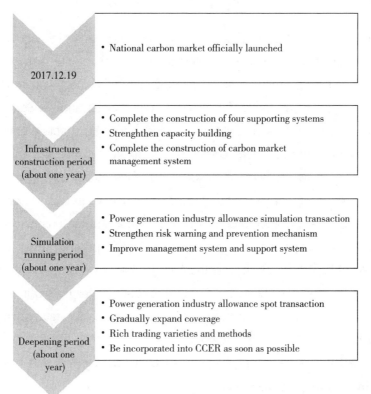

Figure 2 – 1 National carbon market construction stage

The launch of the national carbon market will take the power generation industry as a breakthrough. The reason for this choice is because the historical data of the power generation industry is relatively complete, the products are single, mainly in the heat and electricity categories, and the management is relatively standardized; and the power generation industry has large greenhouse gas emissions. According to the standards of the program, the power generation industry is included in the carbon emission regulatory mechanism. The number of enterprises has reached more than 1700, and the total emissions will exceed 3 billion tons. According to estimates, the carbon emission of the power industry, which is only included in the national carbonmarket, has exceeded the current EU ETS. The China carbon market is bound to become the world's largest carbon emission trading market.

At present, the national carbon market is in the period of infrastructure construction. At this stage, it has been determined that the national market registration system is set up in Hubei, and the transaction settlement system is set up in Shanghai. The Hubei Carbon Emissions Trading Center and the Shanghai Environmental Energy Exchange will take the lead in the infrastructure construction of the relevant systems. The establishment of different systems in the national carbon market in two different provinces and cities mainly takes into account the resource advantages of the two places. The effective indicators such as the number of members of the carbon market in Hubei and the number of participants in the market rank first in the country. The cooperation and exchanges with non-pilot areas are close, and there is a great advantage in membership integration. The status of Shanghai China Financial Center makes it a good financial market environment and the supporting services for establishing a national carbon market are relatively complete.

3. In 2018, there is a big gap between transaction prices and transaction volume in various pilot markets

According to the disclosure of transaction information on each pilot official website, statistics can be obtained that from January to December 2018, the cumulative volume of eight pilots except Sichuan was around 62.42 million tons, and the accumulated transaction amount was about 1.26 billion yuan. Among

II. Progress in China's climate financing

them, the transaction volume and transaction price of each pilot are quite different: the transaction volume and turnover of Guangdong and Shenzhen were relatively large, accounting for 28.362 million tons and 353.4615 million yuan in Guangdong, and 12.6795 million tons and 298.2582 million yuan in Shenzhen. The transaction volume and turnover of Chongqing and Tianjin are relatively small, with Chongqing being 0.2694 million tons and 1.1748 million yuan, and Tianjin being 2.2878 million tons and 26.5484 million yuan. In terms of price, the average price of Beijing allowance is the highest, being about 57.93 yuan/ton; while the average price of Chongqing allowance is the lowest, being about 4.36 yuan/ton, as shown in Figure 2-2, Table 2-2:

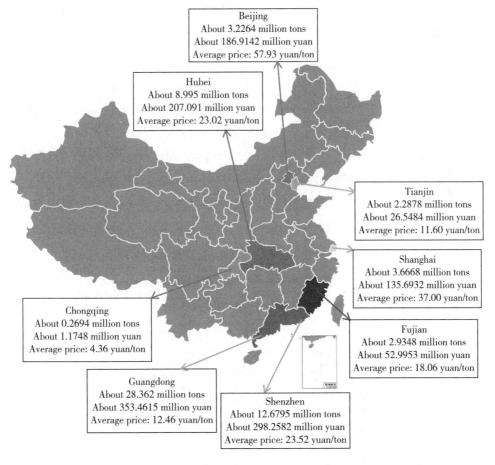

Figure 2-2 Total transaction volume, turnover and average transaction price of each pilot in 2018

Table 2-2   Maximum and Minimum Daily Average Transaction Prices in Pilot Areas in 2018

Pilot	Beijing	Shenzhen	Shanghai	Chongqing	Fujian	Hubei	Guangdong	Tianjin
Date	2018-10-09	2018-06-08	2018-07-30	2018-01-25	2018-11-20	2018-11-22	2018-12-28	2018-06-27
Maximum (CNY/tonne)	74.60	58.91	42.58	31.93	30.00	32.71	18.87	13.96
Date	2018-09-20	2018-12-13	2018-01-10	2018-06-26	2018-12-21	2018-02-01	2018-10-09	2018-06-22
Minimum (CNY/tonne)	30.32	5.24	27.79	2.24	12.00	10.91	1.27	9.24

Data resource: according to the transaction information disclosed on each pilot's website.

(II) The driving force of green finance for climate financing is constantly emerging

Green finance, as a financial solution to promote the green transformation of the world economy, is emerging worldwide. It is forecasted that in the next five years, China needs to invest at least CNY 2 trillion to 4 trillion (USD 320 billion to 640 billion) each year to deal with environmental and climate change issues[1]. Green finance can play an active role in mobilizing funds from all parties to invest in green area. The development of green finance can play a role in stabilizing growth and restructuring the Chinese economy and has risen to a national strategy. Various green financial instruments such as green credit, green bonds, green insurance, and green funds are developing rapidly, mobilizing and stimulating more social capital, while effectively curbing polluting investments and continuing to provide long-term funds for tackling climate change.

1. Green credit

Green credit is the earliest instrument in China's green finance innovation. This concept was first proposed in 2007 by the People's Bank of China, the China Banking Regulatory Commission and other institutions in theOpinions on Implementing Environmental Protection Policies and Regulations

---

[1]   http://wemedia.ifeng.com/67666860/wemedia.shtml.

to Prevent Credit Risks. According to the statistics of the China Banking Regulatory Commission, since the statistics in 2013, the green credit scale of China's 21 major banks grows steadily, with the balance increasing from CNY 4,852.7 billion at the end of June 2013 to CNY 8,295.7 billion at the end of June 2017[①]. Among them, the balance of energy conservation and environmental protection projects and service loans increased from CNY 3,442.9 billion to CNY 6,531.3 billion, and the balance of manufacturing-end loans for strategic emerging industries such as energy conservation, environmental protection, new energy and new energy vehicles increased from CNY 1,423.3 billion to CNY 1,764.4 billion.

As one of the most important channels for green investment and financing in China, green credit plays an important role in mobilizing climate funds. Green transportation and clean energy projects have been the two main areas of green credit investment. At the end of June 2017, the loan balances to the two were CNY 3,015.2 billion and CNY 1,610.3 billion respectively. It is estimated that a total of 179 million tons of standard coal can be saved and 433 million tons of CO_2e can be reduced.

In recent years, China has been vigorously stimulating and guiding the development of green credit. In terms of policy promotion, in December 2017, the People's Bank of China issued the Notice on Promoting the Pledge of Credit Assets and the Internal Bank(Corporate) Ratings, announcing that green loans meet the standards will be given the priority to accept as qualified credit asset collateral for credit policy support re-lending, SLF and other instruments. In January 2018, the People's Bank of China issued

---

① 21 major banks include: China Development Bank, China Exim Bank, China Agricultural Development Bank, Industrial and Commercial Bank of China, Agricultural Bank of China, Bank of China, China Construction Bank, Bank of Communications, China CITIC Bank, China Everbright Bank, Hua Xia Bank, Guangdong Development Bank, Ping An Bank, China Merchants Bank, Pudong Development Bank, Industrial Bank, Minsheng Bank, Evergrowing Bank, Zheshang Bank, Bohai Bank, China Postal Savings Bank. For details, please refer to the official website of the China Banking and Insurance Regulatory Commission: http://www.cbrc.gov.cn/chinese/files/2018/8FF0740BF974482CB442F31711B3ED03.pdf].

the Notice on Establishing a Special Statistical System for Green Loans, clarifying the standards for green loans, and establishing an effective green credit assessment system. In June of the same year, the People's Bank of China announced that it would include quality green loans in the scope of MLF collateral. Subsequently, on 27th July, 2018, the People's Bank of China issued the Notice of the People's Bank of China on the Evaluation of Green Credit Performance of Banking Deposit-Type Financial Institutions and the Green Credit Performance Evaluation Scheme for Banking Deposit-Type Financial Institutions, indicating that the implementation of bank green credit will be directly linked to the results of the Macro Prudential Assessment (MPA). The release of policies can stimulate the development of green credit by banking financial institutions and play a positive role in promoting the entire green credit market.

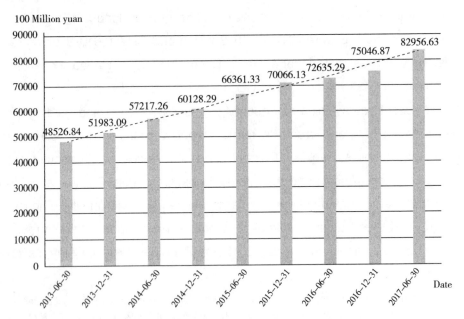

Data Source: China Banking and Insurance Regulatory Commission, http://www.cbrc.gov.cn/chinese/home/docView/96389F3E18E949D3A5B034A3F665F34E.html.

Figure 2-3　Green credit balance of 21 major banks in China

II. Progress in China's climate financing

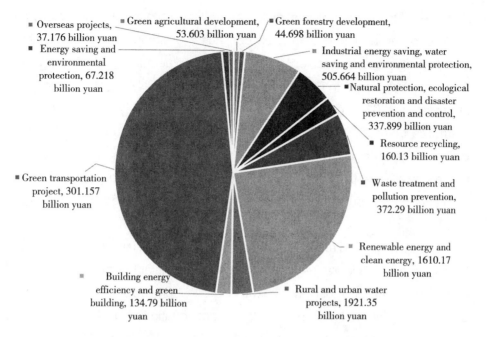

Data Source: China Banking and Insurance Regulatory Commission, http://www.cbrc.gov.cn/chinese/files/2018/8FF0740BF974482CB442F31711B3ED03.pdf.

Figure 2-4 Green credit balance of 21 major banks in China

China Green Credit is actively innovating in product forms, attracting more social capital to the green and climate sectors. Product innovation includes not only the collateralized loans of emerging green equity assets such as future income rights, carbon emission rights and emission rights of contract energy management projects, but also direct investment in green projects with carbon dioxide emission reduction effects. A number of banks launched green credit innovation products in 2018. In June 2018, CCB took the lead in launching a green asset securitization project, signing a contract with five companies for a total of CNY 3.8 billion, and strongly supporting projects such as energy conservation and emission reduction and clean energy utilization to release funds for the mitigation and adaptation of climate change.

2. Green bond

The growth rate of green bond in recent years is remarkable. Under the dual role of favorable policy stimulus and effective financial supervision, the scale of green bonds in China has continued to expand, and both volume and quantity have maintained strong growth momentum. In 2018, the issuance of green bonds in China and abroad reached CNY 267.593 billion, an increase of 8.02% compared with 2017 (CNY 247.714 billion). Among them, 129 domestic green bonds were issued, with ascale of CNY 22.217 billion; 15 green bonds were issued, which was about CNY 45.396 billion. China's total green bond issuance scale is 23.27% globally, down from 2017(24.59%), but it is still one of the world's largest green bond markets[1].

The investment direction of the raised funds for issuing green bonds in 2018 is shown in Figure 2 –5[2]. In addition to financial bonds, clean energy is the area with the most investment in green bond raising funds, totaling CNY 22.998 billion. Followed by clean transportation (CNY 20.25 billion) and resources conservation and recycling (CNY 5.72 billion). Clean energy projects mainly involve wind power generation, solar photovoltaic power generation, and hydropower generation.

Financial institutions have made many favorable attempts in bond innovation, raising funds from the majority of social entities and investing in more climate improvement areas. The innovation of bond varieties has expanded the application of green bonds and increased the market share of green bonds.

---

[1] International Institute of Green Finance, CUFE. 2018 China Green Bond Market Summary [R]. 2018.

[2] The investment direction of the raised funds is classified according to the six categories of funds used in the Green Bond Support Project Catalogue (2015 Edition) issued by the Green Finance Committee (GFC) of China Society of Finance and Banking. Since the use of financial bonds to raise funds needs to be determined based on the subsequent disclosure of information, the "Avatar for multiple purposes" category is set for classification.

II. Progress in China's climate financing

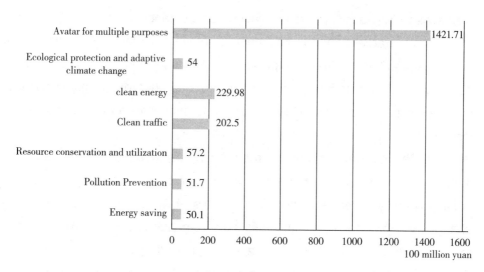

Data Source: International Institute of Green Finance, CUFE. 2018 China Green Bond Market Summary[R]. 2018.

Figure 2 – 5   Use of China Green Bond Raised Funds in 2018

Green bond innovation is not only reflected in the varieties, but also more diversified in the choice of bond buyers. The China Development Bank issued China's first retail green bond for individual investors, which raised the public's awareness of environmental protection and formed a good social effect. China Import and Export Bank issued the first green bond in China for global investors, effectively connecting domestic and foreign capital markets and introducing foreign funds. In addition, Zhejiang Tailong Commercial Bank issued the first green small microfinance bond in China, and raised funds to invest in green credit projects in pollution prevention, resource conservation and recycling, clean energy, ecological protection and climate change adaptation and combined green credit with green bonds.

3. Green insurance

Green insurance gradually exerts its role in ecological governance, green investment and risk management under the promulgation of favorable policies, and is an important climate risk dispersion mechanism. At present, green insurance is being piloted in various provinces and cities in China, and major insurance companies are actively innovating green insurance. In the new energy industry, there are new green insurance innovations such as photovoltaic irradiation index insurance, photovoltaic module performance insurance, and wind power index insurance. In the prevention of natural disasters caused by extreme weather, there are also attempts to manage risk management such as agricultural catastrophe insurance and climate index insurance.

The Chinese government has made many useful attempts in green insurance policy protection, perfecting the system, accelerating legislation, helping green insurance to enhance China's ability to cope with climate change, and preventing climate disasters from bringing economic losses to stakeholders. In May 2017, the China Insurance Regulatory Commission formulated the first environmental protection insurance financial industry standard, Guidelines for Risk Assessment of Chemical Materials and Chemicals Manufacturing Liability Insurance, which provides reference standards for insurance companies' pre-insurance risk assessment. In May 2018, the Ministry of the Environment reviewed and approved the Measures for the Administration of Compulsory Liability Insurance for Environmental Pollution(Draft), which further standardized and improved the system of compulsory liability insurance for environmental pollution. The introduction of relevant policies has promoted the green insurance system to gradually cover areas such as clean energy, ecological restoration, environmental management, ecological agriculture, green buildings and green transportation.

Many provinces and cities in China have carried out innovative green insurance initiatives, including environmental pollution compulsory liability insurance, green liability insurance, and agricultural insurance. In 2017, Shanghai took the lead in launching China's first agricultural typhoon catastrophe index insurance, providing typhoon catastrophe protection for units and individuals engaged in agricultural production and related industries in eight coastal provinces such as Shanghai and Zhejiang. The catastrophe index insurance combining Chinese agricultural characteristics to achieve accurate extreme climate disaster risk assessment has been innovated in three aspects: insurance technology, security coverage, and universal collateral attributes[①].

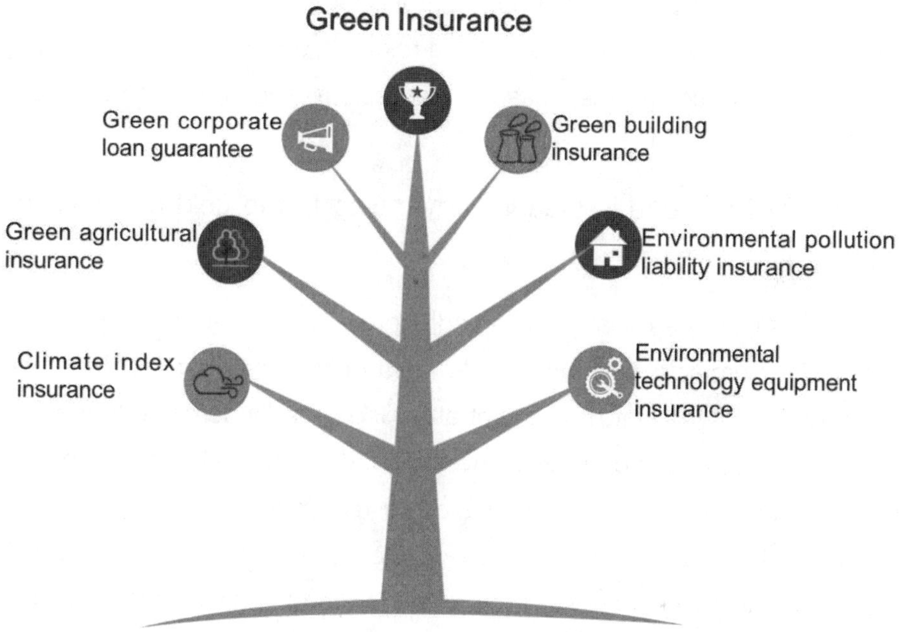

Green Insurance

- Green corporate loan guarantee
- Green building insurance
- Green agricultural insurance
- Environmental pollution liability insurance
- Climate index insurance
- Environmental technology equipment insurance

Figure 2-6 Main classification of green insurance

4. Green fund

The Clean Development Mechanism Fund (CDMF) of the Ministry of Finance is the first climate change policy fund for developing countries. From the full operation in 2010 to 31 December, 2017, CDMF has invested 265 green low-carbon projects through the social fund management model, reducing greenhouse gas emissions 50,136,300 tons of CO_2e[②]. At the local level, the launch of green development funds by local governments in China has become a trend with huge market potential. There are 428 green funds in China, of which about 90% are green industry funds.

---

① Source: http://shanghai.circ.gov.cn/web/site7/tab359/info4077243.htm.
② China Clean Development Mechanism Fund. China Clean Development Mechanism Fund 2017 Annual Report[R/OL]. 2017. http://www.cdmfund.org/zh/jjnb/20609.jhtml.

Green funds are mainly invested in clean energy field. More specifically, green funds flow into new energy vehicles and distributed energy industries. In addition, sewage treatment and environmental restoration projects in the environmental protection industry are also important investment areas for green funds. Green funds can play a role in improving China's ecological environment, reducing greenhouse gas emissions, and enhancing climate adaptability.

(III) PPP mode is steadily advancing in the field of ecological protection

Since 2017, the country has steadily promoted social capital to participate in PPP projects, especially in the field of ecological protection. Under the guidance of the state's policy of attaching great importance to the construction of ecological civilization, investment in ecological construction and environmental protection has increased in recent years. In July 2017, the Ministry of Finance issuedthe Notice on the full implementation of the PPP model for the government-sponsored sewage and garbage treatment projects, stating that the new sewage and garbage treatment projects involving the government's participation will be fully implemented in the PPP model, and establish a unified, standardized and efficient PPP market based on social capital in this field.

According to the data of PPP database of the Ministry of Finance, as of the end of December 2018, there were 8,654 projects in the database, with an investment of CNY13.2 trillion; 365 quarter-on-quarter net increase projects, with an investment of CNY 862.2 billion. The year-on-year net increase of 1,517 projects, the investment amount of CNY 2.4 trillion. At present, there are 4,691 projects (projects already in the transition phase) in the management and transfer phases of the management library (currently 0 projects in the transition phase). The investment amount is CNY 7.2 trillion, and the landing rate is 54.2% (that is, the ratio of the number of projects that have been landed to the number of management library projects).

There are 4,766 projects in pollution prevention and green low-carbon fields, with an investment of CNY 4.7 trillion, accounting for 55.1% and 35.6% of the management pool respectively, and 787 new projects with an annual investment of CNY 0.6 trillion.

In 2017, the PPP reform was progressing steadily, and more emphasis was placed on the normative and high quality of project implementation in the selection process of demonstration projects. Among the third batch of PPP demonstration projects, there are 46 ecological construction and environmental protection projects, mainly based on comprehensive treatment projects. The total investment amounted to CNY 81.056 billion, the number of projects accounted for 8.9% of the total number of projects, and the total investment accounted for 6.9% of the total investment. In 2017, there were 396 demonstration projects in the fourth batch, accounting for 32% of all declared projects, with a total investment of over CNY 758.844 billion. Among them, there are 37 demonstration projects of ecological construction and environmental protection, with a total investment of CNY 55.062 billion, accounting for 9.3% of projects and 7.3% of total investment. Compared with the third batch, as the fourth batch of demonstration projects are more focused on project quality, the amount of investment and the number of projects in ecological construction and environmental protection projects have declined. However, as the total number of projects also decreased by 23.3% and the total investment amount decreased by 35.1%, the proportion of investment in ecological construction and environmental protection projects and the number of projects increased.

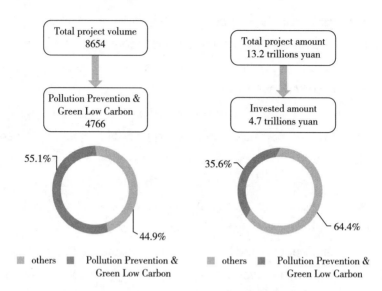

Figure 2-7   2018 Ministry of Finance PPP Project Library Information and Pollution Prevention and Green Low Carbon Project Statistics

In 2017, the national public finance support for energy conservation and environmental protection funds increased significantly compared with 2016, which was CNY 561.733 billion, a year-on-year increase of 18.6%. The support of local financial funds was CNY 526.677 billion, accounting for 3.04% of the total local fiscal expenditure. In terms of specific investment, the national public finance budget for pollution prevention and natural ecological protection was CNY 188.302 billion and CNY 53.710 billion respectively, an increase of 30.1% and 64.5% respectively. Energy management affairs expenditure reached CNY 78.749 billion, an increase of 54.2% over 2016.

II. Progress in China's climate financing

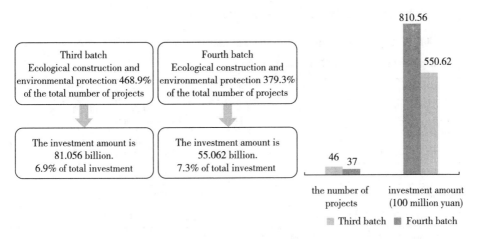

Figure 2 – 8 Statistics on Ecological Construction and Environmental Protection Projects in the Third and Fourth Batch of PPP Projects of the Ministry of Finance

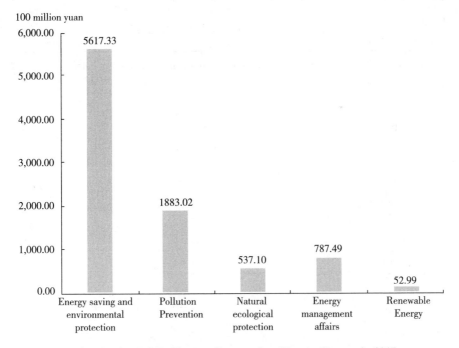

Figure 2 – 9 Public Finance Support for Climate Change in 2017

In 2017, the general public budget expenditure on renewable energy in the country was CNY 5.299 billion, down 38.5% year-on-year, mainly due to the reduction of local budgetary expenditures, and the central government's public budget expenditure in this area also declined.

Table 2 – 3 Comparison of central fiscal and local climate change expenditures with other projects(unit:100 million yuan)

	Central			Local		
Year	2015	2016	2017	2015	2016	2017
Energy saving and environmental protection	400.41	295.49	350.56	4402.48	4439.33	5266.77
Renewable Energy	2.10	10.88	8.71	162.61	75.24	44.28
Education	1358.17	1447.72	1548.39	24913.71	26625.06	28604.79
Science and Technology	2478.39	2686.10	2826.96	3,384.18	3877.86	4440.02
Culture, sports and media	271.99	247.95	270.92	2804.65	2915.13	3121.01
Medical hygiene	84.51	91.16	107.6	11868.67	13067.61	14343.03
Total expenses	80639.66	86804.55	94908.93	150335.62	160351.36	173228.34
Energy saving and environmental protection ratio(%)	0.5	0.34	0.37	2.93	2.77	3.04

Data Source:Financial Data of the Budget Department of the Ministry of Finance, http://yss. mof. gov. cn/2016js/.

II. Progress in China's climate financing

Table 2-4　　PPP Demonstration Project Green Low Carbon Related Industry Statistics*

Field	The first batch of demonstration projects		The second batch of demonstration projects		The third batch of demonstration projects		The fourth batch of demonstration projects		Total			
	Number of items	Investment amount (100 million yuan)	Number of items	Investment amount (100 million yuan)	Number of items	Investment amount (100 million yuan)	Number of items	Investment amount (100 million yuan)	Number of items	Ratio (%)	Investment amount (100 million yuan)	Ratio (%)
1. Transportation:	9	1,565.96	38	3,491.78	40	4467.48	41	2375.99	128	11.15	11,901.21	43.57
Rail	7	1,526.51	13	2,452.67					20	1.74	3,979.18	14.57
Highway			7	613.94	26	3689.36	19	1927.12	52	4.53	6,230.42	22.81
Non-toll road			8	232.38					8	0.70	232.38	0.85
Transportation hub			4	17.32	2	20.57	4	32.14	10	0.87	70.03	0.26
Airport			2	81.98	1	203.17			3	0.26	285.15	1.04
Railway			2	45.7	3	126.41	1		6	0.52	172.11	0.63
Bus			1	15					1	0.09	15.00	0.05
Bridge			1	32.78	4	370.09	2	33.02	7	0.61	435.89	1.60
Other	2	39.45			4	57.87			6	0.52	97.32	0.36
2. Municipal Engineering:	8	77.27	66	810.53	119	1495.35	163	1999.27	356	31.01	4,382.42	16.04
Garbage disposal	1	5.26	22	97.24	31	124.44	21	57.6	75	6.53	284.54	1.04

* Source: Ministry of Finance Government and Social Capital Cooperation Center PPP Project Library.

continued

Field	The first batch of demonstration projects		The second batch of demonstration projects		The third batch of demonstration projects		The fourth batch of demonstration projects		Total			
	Number of items	Investment amount (100 million yuan)	Number of items	Investment amount (100 million yuan)	Number of items	Investment amount (100 million yuan)	Number of items	Investment amount (100 million yuan)	Number of items	Ratio (%)	Investment amount (100 million yuan)	Ratio (%)
Underground integrated pipe gallery	1	13	14	407.77	31	838.68			46	4.01	1,259.45	4.61
Park			1	25	4	43.94	3	31.82	8	0.70	100.76	0.37
Gas supply			2	3.64	2	3.4	1	4.22	5	0.44	11.26	0.04
Heating supply	3	26.4	6	59.52	13	62.24	13	66.87	35	3.05	215.03	0.79
Water supply	3	32.61	18	190.11	24	152.65	12	91.89	57	4.97	467.26	1.71
Sponge city			1	13.85	5	208.9	14	222.55	20	1.74	445.30	1.63
Greening			1	5.5	4	16.22	2	14.13	7	0.61	35.85	0.13
drain			1	7.9	5	44.88	6	31.62	12	1.05	84.40	0.31
3. Environmental protection:	11	111.99	31	742.46	82	932.79	77	922.86	201	17.51	2,710.10	9.92
Sewage treatment	9	59.08	15	337.49	40	181.37	36	508.63	100	8.71	1,086.57	3.98

continued

II. Progress in China's climate financing

Field	The first batch of demonstration projects		The second batch of demonstration projects		The third batch of demonstration projects		The fourth batch of demonstration projects		Total			
	Number of items	Investment amount (100 million yuan)	Number of items	Investment amount (100 million yuan)	Number of items	Investment amount (100 million yuan)	Number of items	Investment amount (100 million yuan)	Number of items	Ratio (%)	Investment amount (100 million yuan)	Ratio (%)
Comprehensive environmental management	2	52.91	15	364.97	38	711.32	29	413.66	84	7.32	1,542.86	5.65
Wetland protection			1	40	4	40.1	2	0.57	7	0.61	80.67	0.30
Total	28	1,755.22	135	5,044.77	241	6895.62	281	5298.12	685	59.67	18,993.73	69.53

(Ⅳ) China actively participates in international exchanges and cooperation, contributing to global control of climate change

China continues to support global climate governance, promote and guide global cooperation on climate change, and actively support developing countries in tackling climate change. China has provided material and equipment assistance to small island states, least developed countries, African countries and other developing countries. It provides strong support for its participation in international negotiations on climate change, policy planning, and personnel training. China also launched cooperation projects in 10 low-carbon demonstration zones, 100 mitigation and adaptation climate change projects and 1,000 training places for climate change in developing countries①.

1. Actively promote the climate negotiation process under the UN framework

In recent years, China is deeply involved in the follow-up negotiations of the Paris Agreement and promotes the establishment of a new global climate governance system. At the 24th UN Climate Change Conference in Katowice, Poland, in December 2018, China actively participated in the negotiation process and promoted the agreement of all parties on key issues. China has made a constructive contribution to the final implementation of the Paris Agreement. In addition, China held the China Climate Investment and Financing forum at China Pavilion to introduce China's achievements in climate finance and share successful experience.

2. Deeply participate in climate change related matters outside the Convention

China voiced at the climate change topics in the Petersburg Climate Change Dialogue and the G20 meeting, the Montreal Protocol, international civil aviation, and international maritime affairs. China continues to pay attention to issues related to climate change under the UN General Assembly,

---

① Source: http://www.china.com.cn/news/2017-10/31/content_41820656.htm.

APEC, BRICS meetings and other occasions. In June 2018, China, the European Union and Canada jointly held the second Ministerial Conference on Climate Action in Brussels, Belgium, to further consolidate the consensus and inject new political impetus in the context of increasing global uncertainty in the climate change process. In September 2018, China, as one of the 17 sponsors, jointly established the Global Commission on Adaptation to promote international cooperation on climate change, accelerate global climate action, and help climate-stricken countries improve climate resilience[1].

3. Commit to promoting climate change South-South cooperation

China has been committed to promoting South-South cooperation on climate change recently. As of April 2018, China has signed a memorandum of understanding with 30 developing countries to provide remote sensing microsatellites, energy-saving lamps, and household solar power systems to help them cope with climate change. China has provided technical assistance in more than 80 developing countries in the areas of clean energy, low-carbon demonstration, agricultural drought-resistant technology, water resource utilization and management, smart grid, green ports, soil and water conservation, and emergency relief and has enhanced the ability of other developing countries to cope with climate change through mitigation and adaptation to climate change projects, donation of energy-saving and low-carbon materials, monitoring and early warning equipment, and organization of training on climate change South-South cooperation[2].

4. Actively promote North-South exchanges and cooperation

China has extensively participated in exchanges between countries and international organizations, actively participated in international conferences, and deepened cooperation with multilateral institutions such as the World Bank, the Asian Development Bank, and the United Nations Develop-

---

[1] Ministry of Ecology and Environment. China 2018 Annual Report on Policies and Actions to Address Climate Change[R]. 2018.

[2] Ministry of Ecology and Environment. 2018 Annual Report on China Policies and Actions to Address Climate Change[R]. 2018.

ment Program. China held a meeting on bilateral cooperation mechanisms on climate change with New Zealand, Germany, France, Canada and other countries and carried out exchanges and cooperation in the fields of carbon market, low-carbon cities, climate change adaptation and so on with the United States, the European Union, France, Germany, the United Kingdom, Canada, Japan and other countries.

II. Progress in China's climate financing

Table 2-5　Development of China's Climate Financing

Classification		Scale	Period(year)	Explanation	Data resource
China's carbon market	China carbon market pilot	In 2018, the annual carbon market allowance reached 62 million tons, with a turnover of about CNY 1.26 billion.	2018	From January to December 2018, the cumulative volume of eight pilots except Sichuan was about 62.42 million tons, and the accumulated transaction amount was about CNY 1.26 billion. Among them, the transaction volume and transaction price of each pilot are quite different; the transaction volume and turnover of Guangdong and Shenzhen are relatively large, and the transaction volume and turnover of Chongqing and Tianjin are relatively small; in terms of price, the average price of Beijing allowance is the highest, and the price of Chongqing is the lowest.	The research team organized the information according to the carbon trading pilot website.
Charitable funds	Donation	Grant from China Green Foundation	CNY 42.57 million	2017	2017 public welfare expenditure CNY 45.41 million	"China Green Foundation 2017 Annual Audit Report"
		The amount of domestic and foreign donations received from China to the ecological environment	CNY 2.07 billion	2017	In 2017, the eco-environment field received a total of CNY 2.07 billion in donations, accounting for 1.38% of the total donation, an increase of 35.3% compared with 2016.	"2017 China Charity Donation Report"

continued

Classification		Scale	Period(year)	Explanation	Data resource
Traditional financial market	Traditional international financial market	China's investment in renewable energy	USD 126.6 billion	2017	China's renewable energy investment is huge, with a year-on-year growth of 30.7% in 2017.	"Global Status Report on Renewable Energy in 2018"
	Domestic financial market	Green credit balance of major banking financial institutions in China	CNY 8.22 trillion	2017-06	—	
		China Banking Industry Energy Saving and Environmental Protection Loan Balance	CNY 6.53 trillion	2017-06	—	Wind information
		China Banking Industry Strategic Emerging Industry Loan Balance	CNY 1.69 trillion	2017-06	—	
		Domestic financial funds, energy saving and environmental protection investment in national public finance	CNY 635.3 billion	2018	—	Ministry of Finance website statistics

II. Progress in China's climate financing

continued

Classification			Scale	Period(year)	Explanation	Data resource
Traditional financial market	Domestic financial market	Labeling the total amount of green bonds	CNY 275.593 billion	2018	—	"China Green Bond Market 2018 Annual Summary"
		Total non-labeled green bond issuance scale	CNY 16150.70 billion	2018	—	
Direct investment of the enterprise	Domestic clean technology field	China's clean technology industry gains VC/PE investment scale	USD 1.21 billion	2018	In 2018, there were 50 financing cases in the clean technology industry, down 20.63% compared with 2017. The financing technology industry's financing scale was USD 1.241 billion this year, a significant decrease of 64.81% compared with the previous year.	Investment in the group's financial data products CVSource
		China's clean technology industry M&A market completed transaction scale	USD 7.969 billion	2018	In 2018, the number of M&A cases completed in the clean technology industry decreased by 29.23% compared with 2017, falling to the lowest level since 2013; the size of completed M&A transactions was USD 7.969 billion, an increase of 17.3% compared with the previous year.	

continued

Classification		Scale	Period (year)	Explanation	Data resource
Direct investment of the enterprise	Domestic clean technology field	China's clean technology companies' domestic IPO financing	USD 1.152 billion	2018	In 2018, a total of 7 Chinese companies in the clean technology industry were successfully listed, mainly in the field of environmental protection and energy conservation. The number of IPO companies fell by 41.67% compared with 2017. In 2018, the clean technology industry raised USD 1,052 million, down 27.04% from the previous year.	Investment in the group's financial data products CVSource
	Warehousing project	Ecological construction and environmental protection	CNY 90 billion	2018	—	
		forestry	CNY 76.5 billion	2018	—	
PPP project	Publish project externally	Pollution Prevention and Green Low Carbon Project	CNY 4.7 trillion	2018	As of the end of December, 4,766 pollution prevention and green low-carbon projects were managed in the reservoir, with an investment of CNY 4.7 trillion. The annual net increase of 787 projects was CNY 1.7 million, with an investment of 0.6 trillion yuan.	Ministry of Finance PPP Project Library

III. Climate investment and financing along the Belt and Road

Control of climate change is a long-term issue that is conducive to the sustainable development of human being. It cannot be effectively solved by a single country and needs global cooperation. In recent years, as the largest developing country, China has not only actively responded to climate change, but also actively provided climate assistance to other countries. In September and October 2013, Chinese President Xi Jinping successively proposedthe Silk Road Economic Belt and the 21st Century Maritime Silk Road, which are in line with the economic development strategy of Asia-Europe economic integration. The human activities in the countries and regions along the Belt and Road are relatively concentrated and strong, and the ecological environment of many countries is fragile, and their ability to adapt to climate change is weak. It is urgent to invest in climate to improve the ecological environment and develop a sustainable economy. Therefore, building a green Belt and Road is an important part of the Belt and Road top-level design.

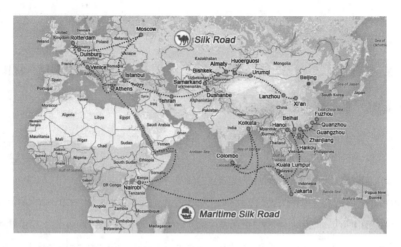

Image source: Xinhuanet.

Figure 3 – 1   Schematic of Beltand Road

(Ⅰ)Background information of the Belt and Road Initiative

1. The Belt and Road Initiative is in line with the development trend of the world

After the global financial crisis in 2008, the recovery of the world economy slowed down, and the deep-seated impact of the financial crisis gradually appeared. The pattern of international investment and trade was to be adjusted, and the development of all countries was not optimistic. In this situation, China proposedthe Belt and Road Initiative in 2013, including the Silk Road Economic Belt and the 21st Century Maritime Silk Road, which is a new concept and initiative for cooperative development. The Belt and Road Initiative adheres to the purposes and principles of the UN Charter and upholds the principle of "communicating, building and sharing" and comprehensively promotes the "five links" principle of policy communication, facility connectivity, trade smoothness, capital finance and people. The Belt and Road Initiative is in line with the development trend of world multi-polarization, economic globalization, cultural diversity and social informatization, and is conducive to establish a global free trade system and an open world e-

III. Climate investment and financing along the Belt and Road

conomy.①.

The Belt and Road Initiative runs through the continents of Asia, Europe and Africa, from the east Asian economic circle to the west of the European economic circle. There are six international economic cooperation corridors along the way, namely: New Asia-Europe Continental Bridge, China-Mongolia-Russia Economic Corridor, China-Central Asia-The West Asia Economic Corridor, the China-Indochina Economic Corridor, the China-Pakistan Economic Corridor, and the Bangladesh-China-Myanmar Economic Cooperation Corridor. They provide an excellent platform for global climate change cooperation②. In the new period of world economic integration and sustainable development, building a large economic corridor for mutual cooperation and common development will bring greater development opportunities to China and countries along the Belt and Road, also enhance the country's economic strength and competitiveness.

2. The construction of the Belt and Road Initiative has made progress

The Belt and Road Initiative has made breakthroughs since launched in 2013. Until the first half of 2018, China has signed cooperation memorandum with 103 countries and international organizations to cooperate in Belt and Road, directly invested more than USD 70 billion in countries along the route, with a 7.2% average annual growth rate, and signed foreign project contracts with countries along the route, summing up to USD 500 billion, with an average annual growth rate of 19.2%③. Data from the Ministry of Commerce shows that till the first half of 2018, China has built 82 overseas eco-

---

① National Development and Reform Commission, Ministry of Foreign Affairs, Ministry of Commerce. Vision and Action to Promote the Construction of the Silk Road Economic Belt and the 21st Century Maritime Silk Road[J]. Zhifu Times, 2015(3):82 −87.

② Liu Zongyi. The Status Quo, Problems and Countermeasures of China's "One Belt, One Road" Initiative in Southeast and Southwest China[J]. Studies in Indian Ocean Economies, 2015(4):92 −109.

③ Source: https://www.sohu.com/a/250863721_498798.

nomic and trade cooperation zones in 24 countries along the Belt and Road, with an investment of USD 2.59 billion. It accounts for 87% of the total new investment in China's overseas economic and trade cooperation zones.

(Ⅱ) Climate characteristics of major countries along the Belt and Road

The Belt and Road Initiative covers 65 countries and regions in Central Asia, West Asia, South Asia, Middle East, Central and South Asia, North Africa, East Africa, Central and Eastern Europe, etc., involving 4.4 billion people along the area, accounting for 63% of the world's total population. The GDP is USD 21 trillion, accounting for 29% of the world's total GDP, covering the world's most dynamic and potential areas of the world[①].

1. Fragile ecological environment

The ecological environment of the countries along the Belt and Road is greatly affected by climate change, and the environmental carrying capacity is poor. The countries in the Central Asian region along the route are mostly deserts, there is lack of green vegetation and water resources, and environmental carrying capacity is very fragile. Environmental pressure in Southeast Asia is also intense. The rapid growth of commercial and industrialization has made the contraction of tropical rain forests and expansion of various industrial pollution. In countries such as West Asia, North Africa, Southeast Asia and South Asia, coastal factories emitted a large amount of pollutants into the ocean, damaging marine ecosystem, endangering biological resources and causing marine pollution. The fragile ecological environment of countries along the route makes it difficult to bear high pollution and high e-

---

① Ding Junfa. "One Belt and One Road" must get through the "five streams" [J]. China Storage and Transportation, 2016(10):46.

mission investment[1].

2. High climate risk

From the aspects of physical impacts of climate change and climate security threats, climate change analysis in five regions of Southeast Asia, South Asia, Central Asia, Northeast Africa and Central Europe along the Belt and Road Initiative can achieve the following results:

Climate change has a great impact on these five regions. The main impact on Southeast Asia is the rise of sea level, the decrease in coastal land and the decline of food production, which further leads to poverty and social instability. The main impact on South Asia is the lack of freshwater resources, the decline of food production, and the poverty problem. Besides, frequent extreme weather causes border conflicts, which influence residents' security. The main impact on Central Asia is water shortages, which makes agricultural production difficult and the increase of potential climate immigrants. What is more, the vegetation in Central Asia is particularly sensitive to precipitation, the drought decreases local biodiversity. The main impact on the Northeast African region is the decline in food production, which leads to violent conflicts caused by the looting of resources, and the spread of extremism and terrorism. The main impacts on Central Europe are water shortages and frequent extreme weather, affecting residents' safety and leading to resource contention[2].

---

[1] Yang Zhen, Shen Enwei. Discussion on Accelerating Green Investment in Countries along the Line under the Belt and Road Strategy[J]. Foreign Economic Relations and Trade, 2016(9): 21-24.

[2] Wang Zhifang. Climate Security Risks in China's Construction of the Belt and Road [J]. International Political Research, 2015, 36(4): 56-72.

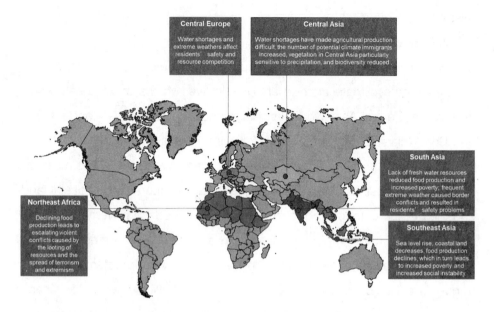

Figure 3-2　Main climate risk map along the Belt and Road

3. The total amount of emissions is huge, and the potential for emission reduction is considerable

China is the world's largest emitter of carbon dioxide. Many countries along the Belt and Road Initiative are ecologically fragile and sensitive. Asian countries have become the world's fastest growing fossil energy consumption parties. As shown in the table below, the carbon emission of the countries along the Belt and Road Initiative in 2015-2017 are generally on the rise, and the Belt and Road Initiative region covers the world's major carbon emitters, such as China, India, Russia. In 2017, the total carbon emission of countries along the Belt and Road was about 22.758 billion tons of CO_2e, accounting for 63% of global total carbon emission. The economic development pattern of these countries is relatively extensive, and the production and operation activities bring huge burdens to the local ecological environment, environmental problems are prominent and pollution is serious. It is urgent to adjust the economic development pattern and carry out low-carbon investment. If effective measures are taken, there will be great potential

III. Climate investment and financing along the Belt and Road

for emission reduction in these countries.

Table 3 – 1 Total carbon emissions in 2015 – 2017 for major countries along the Belt and Road (unit: million tons of CO_2e)

Region	Country	2015	2016	2017
12 countries in East Asia	China	9,975.3554	9,967.9553	10,110.279
	Malaysia	245.0976	251.0896	254.5759
	Singapore	60.7206	62.8291	64.7651
	Vietnam	184.5019	197.8016	198.8265
	Myanmar	23.4407	24.0176	25.3332
	Cambodia	6.9765	7.482	7.9383
	Laos	1.745	1.8985	1.958
	Indonesia	459.6288	464.8568	486.8438
	Brunei	10.3553	9.9647	10.2269
	Philippines	113.8232	119.2737	127.608
	Mongolia	20.3564	28.681	30.3907
	Thailand	320.574	324.8341	330.8396
8 countries in South Asia	Pakistan	172.09	187.4054	198.81
	Afghanistan	10.1074	12.2579	13.0147
	Bengal	83.5008	85.2536	88.0575
	Sri Lanka	20.7134	21.9507	23.1384
	Maldives	1.4554	1.4721	1.5718
	Nepal	7.5854	8.526	9.0282
	India	2,276.4072	2,377.4479	2,466.7654
	Bhutan	0.98796	1.1146	1.1664
West Asia 18 countries	Saudi Arabia	620.883	631.5492	635.0111
	Iraq	167.7119	184.9387	194.4546
	Iran	630.3618	637.5622	672.3123
	Oman	64.2211	64.4421	65.1864
	Qatar	120.5472	118.6898	129.8033
	Kuwait	102.327	103.456	104.3935
	Lebanon	21.2692	19.7574	19.5468

continued

Region	Country	2015	2016	2017
West Asia 18 countries	Turkey	380.8581	402.8208	447.8972
	Egypt	200.4222	208.9113	218.6644
	Syria	28.9252	27.9994	27.9145
	Jordan	23.1678	21.6301	21.3567
	Israel	66.0959	65.1682	66.5534
	Yemen	20.3846	19.1354	18.9658
	United Arab Emirates	227.4772	233.4265	231.7735
	Bahrain	32.8469	34.2651	34.4557
	Cyprus	6.9029	7.3078	7.5147
	Greece	74.9625	71.3731	76.0004
	Palestine	2.5541	2.3405	2.318
5 countries in Central Asia	Turkmenistan	75.379	76.5633	72.7025
	Kazakhstan	263.4839	278.3619	292.5885
	Kyrgyzstan	9.9294	9.8599	10.4331
	Tajikistan	5.4602	5.579	5.7114
	Uzbekistan	110.6693	97.9253	98.9989
CIS 7	Azerbaijan	40.2414	39.6883	38.2097
	Armenia	5.7917	5.9608	5.6564
	Belarus	58.9658	60.5092	61.3718
	Moldova	4.9487	5.2004	5.0936
	Ukraine	223.5797	235.156	212.1163
	Georgia	9.4636	9.8389	10.9517
	Russia	1671.8951	1668.0699	1692.7948
16 countries in Central and Eastern Europe	Lithuania	13.1416	13.1575	13.3939
	Latvia	7.3337	7.2636	7.1668
	Estonia	15.8911	17.4935	19.8093
	Bosnia	24.7642	26.0017	26.6469
	Albania	5.9477	6.2729	6.3792

III. Climate investment and financing along the Belt and Road

continued

Region	Country	2015	2016	2017
16 countries in Central and Eastern Europe	Slovenia	13.5991	14.3998	14.6096
	Croatia	17.9966	18.2215	17.1816
	Serbia	41.3375	43.351	45.0928
	Macedonia	7.2575	7.144	7.2515
	Czech Republic	104.7846	106.5433	107.8958
	Slovakia	33.8975	33.9968	35.3862
	Bulgaria	48.1327	45.2874	49.0712
	Hungary	46.6652	47.5782	50.3447
	Poland	310.6151	322.234	326.6045
	Romania	77.7882	75.0517	79.9955
	Montenegro	2.4392	2.5735	2.6278
Total		22,053.74086	22,306.1691	22,758.3457
Global		35,462.7467	35,675.0994	36,153.2616

(III) The foundation to promote green investment along the Belt and Road

Most countries along the Belt and Road have weak economic foundations and serious environmental problems, therefore in the process of economic construction, the improvement of the ecological environment is a very important consideration. The countries along the line have low degree of openness to the outside world, low level of social and economic development, and lack of climate investment concept. China actively takes the responsibility of major power in the construction of the Belt and Road, guides Chinese enterprises to make green investments in the local area, proposes to establish a green strategic alliance, and shares relevant experience of environmental governance and climate improvement to relevant parties involved in the Belt and Road construction, maintains a balance of economic, environmental and social benefits in investment.

1. Strengthening the green concept: the Belt and Road green development strategy framework

In order to solve the problems encountered in the Belt and Road construction process, the Ministry of Environmental Protection proposed the Belt and Road green development strategy framework in 2016. The framework is proposed to carry out green industrial output, investment construction, technical assistance cooperation, strengthen ecological and environmental protection cooperation, and avoid ecological environmental risks in the Belt and Road construction. The Belt and Road involves a lot of countries, they have large population and huge greenhouse gas emissions. If the extensive economic development pattern is still adopted, the goals set by the Paris Agreement will be difficult to achieve.

The strategic framework is designed in four levels. First is to strengthen the top-level design for strategy implementation, implement green development planning, especially to do eco-environmental special planning. Second, establishing an early warning mechanism for ecological environment risks, accurately identify and circumvent the ecological environment risks of cooperation with countries alongthe Belt and Road. Third, strengthening the regulation of foreign investment cooperation, motivate environmental investment in foreign investment and cooperation activities. Fourth is to strengthen the construction of common support capabilities for the implementation of the Belt and Road green development strategy, establish a sound civil exchange mechanism, a detailed publicity and advertising mechanism, a developed information platform network, to support the implementation of the green development strategy. The specific strategic content is shown in the figure below.

III. Climate investment and financing along the Belt and Road

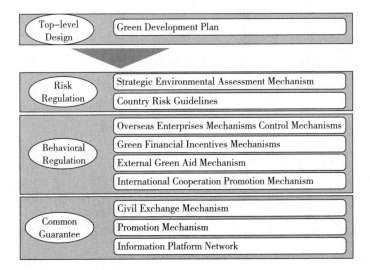

Data Source: Dong Zhanfeng, Ge Chazhong, Wang Jinnan, etc. *Strategic Implementation Framework for Green Belt Development of " One Belt, One Road"* [J]. China Environmental Management, 2016,8(2):31 -35.

Figure 3 -3 The Belt and Road Green Development Strategy Framework

2. Strengthening green international cooperation: The Belt and Road Green Development International Alliance

Building a green finance system, greening the financial system, and developing a green economy are not only a new trend in global economic development, but also a requirement for China to strengthen environmental protection and ecological civilization construction. It is also an important development direction for the future construction of the Belt and Road.

On 14th May 2017, President Xi Jinping attended the Belt and Road International Cooperation Summit Forum, he raised four proposals, including establishing of the Belt and Road International Development International Alliance, strengthening the innovation cooperation among countries along the Belt and Road, launching the Belt and Road Science and Technology Innovation Action Plan, and carrying out scientific and humanities exchanges, joint laboratories construction, cooperation in science and technology parks, and technology transfer. These proposals are highly concerned by the United Nations and all society sectors. The alliance is committed to promoting cooperation in environmental protection, ecological construction and financial development in countries along the Belt and Road with the help of an international platform. The establishment of the alliance will help strengthen South-South cooperation, North-South cooperation, multilateral cooperation and multi-party cooperation under new forms, and will attract more business institutions, green funds, climate funds, non-governmental organizations (NGOs) and other groups to carry out environmental protection funding and technical assistance[①]. At present, the United Nations Environment Program and the Ministry of Ecology and Environment are leading the establishment of the alliance, unite international organizations, enterprises, financial institutions, think tanks and civil organizations to start the alliance work as soon as possible.

3. Sharing green governance experience: Belt and Road Desert Green Economy Innovation Center

On 24th June 2017, the United Nations Environment Program and the Yili Public Welfare Foundation jointly established the Belt and Road Desert Green Economy Innovation Center in the Kubuqi Desert, Erdos City, Inner Mongolia. The establishment of the center aims to build a communication platform to share Kubuqi's more than 30 years' experience in desert green economy development in global desertification areas, and promote Kubuqi's sand control experience, achievements and technology. This is an important mechanism to promote the international cooperation and technological innovation of the Belt and Road green development.

The center has three operation platforms, namely Belt and Road ecological technology research and development platform, technology achievement transfer and application platform, and youth environmental education platform. The contents of the center work are: carrying out scientific and

---

① Source: http://jjckb.xinhuanet.com/2017-07/03/c_136412229.htm.

technological innovation to promote basic research on ecological restoration; cultivating professionals in ecological restoration and environmental management; organizing environmental protection training and outreach activities to enhance teenagers' awareness of environmental protection; carrying out multilateral and bilateral cooperation on ecological restoration and green development along the Belt and Road area; promoting the integration of industry-university research and industrialization of scientific and technological achievements; and share the Chinese experience of desertification control, promote international cooperation in ecological governance and combating desertification along the Belt and Road①.

4. Convergence Green Investment Consensus: Belt and Road Green Investment Principle

In order to promote the green investment of Belt and Road Initiative, the China Finance Association Green Finance Committee and the London Financial City Green Finance Initiative jointly issued the Belt and Road Green Investment Principles on 30th November 2018. Based on the existing responsible investment principle, this principle strengthens the concept of low carbon and sustainable development and improves the investment environment and social risk management level.

To ensure that investment projects have a positive external impact on climate and environment, and to support the implementation of the Paris Agreement, the principles include seven suggestions, including integrating sustainability into corporate governance, fully understanding ESG risks, fully disclosing environmental information, strengthening communication with stakeholders, making full use of green financial instruments, adopting green supply chain management, and building capacity through multi-party cooperation.

The Belt and Road Initiative construction contains huge opportunities

---

① Source: http://www.xinhuanet.com/energy/2017-06/26/c_1121212487.htm.

for climate investment. The establishment of a green governance platform will enable countries along the initiative to establish a community to deal with environmental issues and improve their environmental risk response capabilities. If Belt and Road Initiative can help the countries to develop a low-carbon economy through the new concept of green output and low-carbon development, it can not only reflect the purpose of equality and mutual benefit, win-win cooperation and common development of the Belt and Road Initiative, but also promote the sustainable economic and social development of China and countries along the initiative.

(Ⅳ) China's climate investment and financing projects for countries along Belt and Road Initiative

Since the introduction of Belt and Road Initiative, Chinese companies have actively taken part in the project construction along the initiative, enhancing countries' ability to mitigate and adapt to climate change. The climate investment and financing in Belt and Road Initiative has large volume, large number of participating institutions and various investment methods.

Due to the high climate risk, low socio-economic development level and poor infrastructure construction along the Belt and Road Initiative, China's climate investment is mainly concentrated on clean energy and clean transportation. On the one hand, these investments can exert the "multiplier effect" of infrastructure, solve the sustainable and stable development problem for most countries along the route, and improve the overall living standards of citizens; on the other hand, the environmental benefits of sustainable infrastructure are huge. For example, Nirum-Jellum, the largest hydropower project in Pakistan invested by China, can generate an annual power generation of about 5.15 billion kWh after the completion of all the power generation units, which accounts for 12% of Pakistan's hydropower generation and can solve the problem of power shortage in 15% of Pakistan's national pop-

III. Climate investment and financing along the Belt and Road

ulation. The emission reduction effect is obvious[①].

As of the end of October 2018, there were 51 clean transportation and clean energy projects in China along the Belt and Road Initiative, including 20 clean transportation projects and 31 clean energy projects (see Table 3-2). The project covers all major continents such as Southeast Asia, South Asia, West Asia, Central Asia, Europe and Africa. These projects not only meet the needs of local residents, improve overall living standards, but also improve local climatic conditions.

Case 1

China's construction of first light rail in East Africa

On 20th September 2015, the first light rail in East Africa which was built by China, was opened for trial operation in Addis Ababa, Ethiopia. The Addis Ababa Light Rail Project has a total length of 32 kilometers, including two lines, Line 1 and Line 2, running through the city from east to west and north to south. The light rail runs for 16 hours, 244 times a day in Addis Ababa. The daily average passenger capacity is about 100,000, which will effectively ease the traffic congestion in Addis Ababa.

The project has a total investment of USD 475 million, of which 85% is provided by the Export-Import Bank of China and 15% by the local government. The current light rail project is the first phase of the entire light rail project in Ethiopia and is expected to reduce 10% of urban traffic pressure. The second phase of the project is currently waiting for follow-up funding.

---

① Source: https://www.yidaiyilu.gov.cn/xwzx/hwxw/52755.htm.

> The operation of the light rail project will effectively reduce local carbon emissions. Before the light rail was put into operation, urban traffic emissions accounted for most of the total greenhouse gas emissions in Addis Ababa. The 2012 Addis Ababa Greenhouse Gas Emissions Inventory shows that 47% of the 4.8 million t CO_2e emitted in the city comes from transportation. The light rail has been operating safely for 985 days from the second half of 2015 to 31st May 2018, with a total of 230,000 trains running, a mileage of 5.842 million kilometers, a total carriage of 129 million passengers and a daily average of 10,450 passengers. The highest single-day passenger flow in history was 185,000[①]. Light rail operations have effectively reduced greenhouse gas emissions in Addis Ababa. According to Bloomberg Philanthropy and China Greentech developer BYD, it is expected that by 2030, the Addis ababa light rail project will reduce the city's emissions by 1.8 million tCO_2e. In 2016, Addis Ababa was awarded the 2016 C40 City Award, which demonstrated its efforts to address climate change on a global scale, and the Light Rail project is an important reason for its success. The award recognizes the transformation of Addis Ababa from diesel fuel public transport to more eco-friendly transport.

Table 3-2 China's climate investment and financing projects along the Belt and Road

No.	Category	Country of project	Project name
1	Railway traffic	Laos	China-Laos Railway Project
2	Railway traffic	Kenya	Kenya Mombasa to Nairobi Railway Project

---

① Source: http://www.sohu.com/a/233930118_157267.

III. Climate investment and financing along the Belt and Road

continued

No.	Category	Country of project	Project name
3	Railway traffic	Belarus	Belarusian Railway Electrification Project
4	Railway traffic	Russia	Moscow-Kazan High Speed Rail Project
5	Railway traffic	Myanmar	Sino-Myanmar Wood Mister-Mandalay Railway Project
6	Railway traffic	Myanmar	China-Myanmar International Railway Project
7	Railway traffic	Bangladesh	Bangladesh Dhaka to Chittagong High Speed Railway Project
8	Railway traffic	Thailand	Zhongtai Railway
9	Railway traffic	Thailand	Bangkok Mass Rapid Transit Project (Pink Line and Yellow Line)
10	Railway traffic	Hungary, Serbia	Hungarian Railway
11	Railway traffic	Indonesia	Yawan High Speed Rail Project
12	Railway traffic	Ethiopia, Djibouti	Addis Ababa-Djibouti Railway
13	Railway traffic	Pakistan	Pakistan Lahore Rail Traffic Orange Line Project
14	Railway traffic	Saudi Arabia	Mecca-Medina high speed railway project
15	Urban rail transit	Israel	Red Line Light Rail Project
16	Urban rail transit	Georgia	Georgia Modern Railway Project
17	Urban rail transit	Egypt	Egyptian Suburban Railway Project on the 10th of Ramadan
18	Urban rail transit	Bangladesh	Dhaka Urban Rail Transit Project
19	Urban rail transit	India	Mumbai Metro Line 1
20	Waterway traffic	Bangladesh	Padma Bridge and River Dredging Project
21	Wind power	Ethiopia	EthiopianAdama Wind Power Project
22	Wind power	Malta	MontenegroMozula Wind Power Project

continued

No.	Category	Country of project	Project name
23	Wind power	Pakistan	Jimpur Wind Power Project
24	Solar photovoltaic power generation	Pakistan	900 MW photovoltaic ground power station project in Punjab, Pakistan
25	Solar photovoltaic power generation	Pakistan	Pakistan 100 MW large solar photovoltaic power station
26	Solar photovoltaic power generation	Algeria	Algeria 233 Photovoltaic Power Plant Project
27	Solar photovoltaic power generation	Eritrea	China aids Eritrea solar photovoltaic project
28	Solar photovoltaic power generation	Hungary	Greensle Solar Power Station
29	Solar photovoltaic power generation	Argentina	Gaocharay Photovoltaic Power Station
30	Solar photovoltaic power generation	Ukraine	N/A
31	Solar thermal + solar photovoltaic power generation comprehensive project	Morocco	Nuo Ouzazazate Solar Thermal Power Generation Complex
32	Natural gas project	Myanmar	YangonDajida Natural Gas Combined Cycle Power Plant Project
33	Hydropower project	Laos	Laos South Russia 3 Hydropower Project
34	Hydropower project	Laos	LaosHueland Pangre River Hydropower Project
35	Hydropower project	Laos	LaosNanxun Hydropower Station
36	Hydropower project	Pakistan	PakistanKarlot Hydropower Project
37	Hydropower project	Pakistan	Nirum-Jelum Hydropower Station
38	Hydropower project	Cote d'Ivoire	Subray Hydropower Project
39	Hydropower project	Nepal	Shangma Xiangdi A Hydropower Station
40	Hydropower project	Ethiopia	Gib 3 Hydropower Station

III. Climate investment and financing along the Belt and Road

continued

No.	Category	Country of project	Project name
41	Hydropower project	Ethiopia	Tekze Hydropower Station
42	Hydropower project	Angola	Kakolu Kabassa Hydropower Project
43	Hydropower project	Ecuador	EcuadorCocacodo-Sinclair Hydropower Station
44	Hydropower project	Cambodia	Cambodia Elsie Hydropower Station
45	Hydropower project	Uganda	Karuma Hydropower Station
46	Hydropower project	Guinea	Kaile Tower Hydropower Station, Guinea
47	Hydropower project	Cameroon	Cameroon Manville Hydropower Station
48	Hydropower project	Malaysia	Malaysia Baler Hydropower Station
49	Hydropower project	Belarus	Vitebsk Hydropower Station
50	Geothermal project	Indonesia	Indonesia SMGP 240MW Geothermal Power Project
51	Nuclear power project	Pakistan	Pakistan Nuclear Power Plant Project

Source: China Belt and Road Network: https://www.yidaiyilu.gov.cn/info/iList.jsp?cat_id = 10005, Asian Investment Bank official website: https://www.aiib.org/en/index.html.

(V) The main institutions for the Belt and Road climate investment and financing

Since the implementation of the Belt and Road Initiative, Chinese financial institutions have actively participated in project construction and provided financial support for enterprises. It is estimated that the infrastructure investment gap in the Belt and Road region will exceed 600 billion USD per year[1]. If potential climate risks are considered in the early stage of construction and climate resilient infrastructure is built, the environmental benefits will be enormous and sustainable. The main institutions involved in the

---

[1] Source: https://baijiahao.baidu.com/s?id = 1597522887730681238&wfr = spider&for = pc.

construction of the Belt and Road include Chinese banks and international financial institutions initiated by China.

1. Chinese banks are actively investing in climate investment and financing along the Belt and Road

(1) China Development Bank

In 2017, China Development Bank (CDB) completed 22 major international plans for the Belt and Road and increased loans of USD 17.6 billion in the countries along the route. CDB finances the infrastructure interconnection, capacity and equipment manufacturing cooperation, financial cooperation and overseas industrial park construction along the region. In addition, CDB has committed a total of CNY 99.1 billion in credit grants for CNY special loans, initiated the establishment of the China-Central and Eastern European Union Bank, and promoted multi-bilateral financial cooperation with the Shanghai Union Bank, China-ASEAN Union Bank, and the BRICS Bank Cooperation Mechanism[①].

In 2017, CDB successfully issued the first USD 500 million and 1 billion euros of China's quasi-sovereign international green bond. The raised fund supports green industry projects related to clean transportation, renewable energy and water resources protection in Belt and Road, aims to improve the ecological environment of countries along the route and enhance the ability of countries to deal with climate change. In addition, CDB issued a USD 350 million Belt and Road special bond in Hong Kong through private placement, and innovatively interconnected the mainland and Hong Kong markets to support the Belt and Road financing pattern. Leading the underwriting of Malayan Bank's 1 billion bond-to-bond CNY panda bond, which is specially used to support the construction of Belt and Road project at home and abroad. It is the first bond-to-bond panda bond in the ASEAN countries

---

① National Development Bank. Annual Report of the National Development Bank 2017 [R]. 2017.

III. Climate investment and financing along the Belt and Road

as well as in China bond market①.

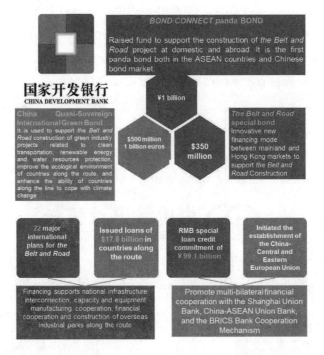

Figure 3-4 China Development Bank's Belt and Road Climate Investment and Financing Situation

(2) China Export-Import Bank

In 2016, the China Export-Import Bank invested in the Kalot Hydropower Project in Pakistan. The planned installed capacity of the project is 720,000 kilowatts, with a total investment of about USD 1.65 billion. It is expected to be put into operation in 2020. After completion, it will provide Pakistan with about 3.2 billion kWh of clean energy every year, which will effectively alleviate the shortage of electricity in Pakistan.

On 22nd December 2017, China Export-Import Bank's first bond-to-bond green bond was issued in Shanghai for a period of three years. The bond

---

① National Development Bank. Sustainable Development Report of National Development Bank 2017[R]. 2017.

had a value of CNY 2 billion and an issue rate of 4.68%. A number of banks including Bank of China Singapore Branch and Hong Kong, Europe and other overseas investment institutions actively participated in the issuance of subscriptions, participating in the subscription amount of CNY 520 million, and the final placement amount of CNY 260 million. The green bond fund will be invested in clean energy and environmental improvement projects along the Belt and Road. The independent third-party assessment estimated that the fundraising will achieve good environmental benefits in reducing carbon dioxide, sulfur dioxide and nitrogen oxides[①].

Figure 3 – 5　China Export-Import Bank's Belt and Road Climate Investment and Financing Situation

(3) Bank of China

Bank of China (BOC) has branches in 23 countries along the Belt and Road, which has the largest number of branches along the region among Chinese banks. As of the end of 2017, BOC has followed up more than 500 major projects in the Belt and Road. Between 2015 and 2017, about USD

---

① The import and Export Bank of China, the "green bonds" of the import and Export Bank of Shanghai, has successfully issued the green debt, helping to build the green "one belt and one road" global interconnection. http://www.eximbank.gov.cn/tm/Newlist/index_343_30615.html.

III. Climate investment and financing along the Belt and Road

100 billion in credit support was provided to countries along the Belt and Road. BOC also assisted the Hungarian government in issuing the first sovereign panda bond to raise funds for the Belt and Road Initiative[①].

Figure 3-6 Bank of China's Belt and Road Climate Investment and Financing Situation

(4) Industrial and Commercial Bank of China

The Industrial and Commercial Bank of China (ICBC) Luxembourg Branch issued the first Belt and Road climate bond with an issue value of USD 2.15 billion, which was oversubscribed by global investors. The bond is issued in three instalments and covers both the US dollar and the euro. Bond proceeds will be invested in renewable energy, low carbon and low emission transportation. By the end of 2017, ICBC has accumulatively supported 358 Belt and Road projects, with a total contracted loan amount of USD 94.5 billion, and 123 new loan-to-finance projects in 2017, with a loan amount of USD 33.9 billion. ICBC has 129 branches in 20 BRI countries and regions[②].

---

① Bank of China. Bank of China Annual Report 2017[R]. 2017.
② Industrial and Commercial Bank of China. Industrial and Commercial Bank of China Annual Report 2017[R]. 2017.

Figure 3 – 7　Industrial and Commercial Bank of China's Belt and Road Climate Investment and Financing Situation

2. The financial institutions initiated by China bear the important role of climate investment and financing along the Belt and Road

The Asian Infrastructure Investment Bank(AIIB)and the Silk Road Fund are the main sources of funding for the Belt and Road Initiative. The two institutions differ in their investment methods. The AIIB focuses on debt investment,mainly through the issuance of loans to participate in the project, while the Silk Road Fund operating model is biased towards direct equity financing,and the investment period is longer. The two can also generate more fundraising methods to raise more funds for countries along the Belt and Road[①]. At the same time,the funds of AIIB is mainly funded by governments,which is government behavior. The Silk Road Fund is mainly for entities that have funds and want to invest,which means that Silk Road Fund can absorb private capital to participate in the Belt and Road construction.

(1)Asian Infrastructure Investment Bank

The Asian Infrastructure Investment Bank(AIIB)is the first multilateral financial institution set up by China to support infrastructure construction. As

---

① 　Source:http://opinion. hexun. com/2017 −05 −20/189268794. html.

III. Climate investment and financing along the Belt and Road

of December 2018, the AIIB has 87 member states or candidate countries, of which 44 are within the region that have subscribed for equity and 24 are members from outside the region[①].

The AIIB is an important source of funding for the Belt and Road Initiative. As of December 2018, the AIIB official website listed a total of 31 approved projects involving a financing amount of USD 6.295 billion, of which 19 climate-related investment projects, financing amount of USD 4.411 billion. The number of projects accounts for more than 60% of the total investment projects, and the financing amount accounts for more than 70% of the total financing amount. The projects involve renewable energy, green transportation, urban waste treatment, sewage treatment and so on.

In 2018, AIIB invested eight investment projects in the countries along the Belt and Road. Except for the National Investment and Infrastructure Fund project in India, which was funded in FoF to attract institutional investors'private capital to reduce the equity financing gap in India's infrastructure sector, the remaining 7 projects are invested in climate mitigation and adaptation projects, including 3 clean energy projects, 2 water resources projects and 2 clean transportation projects.

Analysis from the project location shows that 3 of the new projects are invested in India with a total investment of USD 695 million, 2 projects are invested in Turkey with a total investment of USD 800 million, Egypt, Indonesia and Bangladesh respectively has 1 project, the investment amount is USD 300 million, USD 250 million, and USD 0.6 billion for each.

[①] Source: https://www.aiib.org/en/about – aiib/governance/members – of – bank/index.html.

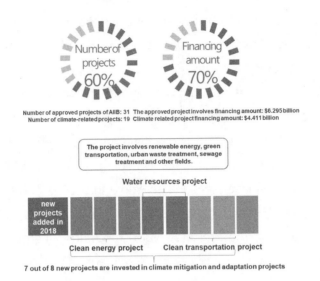

Figure 3 – 8 Climate Investment and Financing Projects of Asian Infrastructure Investment Bank

(2) Silk Road Fund

Silk Road Fund was established in Beijing on 29th December 2014. It focuses on the construction of infrastructure, resource development, capacity cooperation and financial cooperation in the relevant countries and regions around the construction of the Belt and Road. Silk Road Fund mainly focuses on equity investment, and combines various methods such as bond, fund and loan to provide investment and financing services for the Belt and Road construction.

The Silk Road Fund has a fund size of USD 40 billion and CNY 100 billion. Among them, foreign exchange reserves (through Wutongshu Investment Platform Co., Ltd.), China Investment Corporation (through Cyrus Investment Co., Ltd.), China Export-import Bank, and China Development Bank (Guokai Financial Co., Ltd.) hold the share of 65%, 15%, 15% and 5% respectively[1].

---

[1] Source: Silk Road Fund Official Website, http://www.silkroadfund.com.cn/cnweb/19854/19858/index.html.

III. Climate investment and financing along the Belt and Road

Since its establishment in 2014, Silk Road Fund has actively participated in the construction of projects along the Belt and Road. The project has a wide coverage and various types, including infrastructure, resource development, capacity cooperation and financial cooperation. Silk Road Fund has tracked and stored more than 100 projects along the Belt and Road, covering Russia, Mongolia, Central Asia, Southeast Asia, South Asia, West Asia and North Africa, Central and Eastern Europe and other key regions and countries. ①

Figure 3-9 Investors' Investment Ratio of Silk Road Fund

From the investment projects of Silk Road Fund, the existing climate investment is mainly focus on clean energy, including hydropower, natural gas, clean coal-fired power generation, and photovoltaic power generation. In terms of investment methods, Silk Road Fund tends to participate in projects through equity investment. By the end of March 2018, more than 70% of the investment in Silk Road Fund was through equity investment.

For some large scale projects, Silk Road Fund also provides loans at the same time as the share subscription, this innovative model of equity + credit investment achieved a win-win situation for enterprises and Silk Road Fund. On the one hand, equity investment reduces the asset-liability ratio of the project, making it easier for the project to obtain financial support. On the other hand, debt investment can generate relatively safe and stable earnings, which reduces the risk of sole equity investment. The equity +

---

① Source:http://www.financialnews.com.cn/zgjrj/201704/t20170428_116703.html.

credit model makes a balance between the risks and benefits of investment. Silk Road Fund has adopted equity+credit model in the Karat Hydropower Project in Pakistan, the Yamal LNG Integration Project in Russia, and the Harbin Clean Coal-fired Power Plant Project in Dubai.

In the process of project construction, Silk Road Fund also pays attention to theprotection of environment where the project is located, and minimizes the negative impacts of the project on local ecosystems and species. The projects adopted the most advanced and strict production technology standards, met local carbon emission standards and took corresponding protective measures in the construction process.

Case 2

Silk Road Fund invests in the first clean coal-fired power station in the Middle East

The Dubai Haxiang Clean Coal-fired Power Station is located in more than 30 kilometers south of Dubai, United Arab Emirates. This is the first clean coal-fired power station in the Middle East under construction, and also the first power station project invested, constructed and operated by Chinese enterprises in the Middle East under the framework of Belt and Road. The Dubai Haxiang clean coal-fired power station project is a key project of the Dubai government's energy strategic plan. The power station consists of four 600MW units with a total investment of about USD 3.3 billion. The first unit is scheduled to be put into commercial operation in March 2020. It is expected to meet 20% of Dubai's total electricity demand after full operation in 2023.

As a project investor, Silk Road Fund not only provided loans itself, but also promoted the participation of Chinese financial institutions in syndicated loans, which led to a total of about USD 1.6 billion of Chinese bank credit. Silk Road Fund adopts the equity+credit model. On the one hand, equity investment reduces the asset-liability ratio of the project, making it easier for the project to obtain funding support; on the other hand, debt investment and

III. Climate investment and financing along the Belt and Road

equity investment form a complement and cooperation, taking into account the balance between investment risk and return.

In terms of production technology, the Haxiang project uses the most advanced combustion, denitration, dust removal and desulfurization technologies to ensure that dust, sulfide and nitride emissions emitted during power plant operation are superior to those of the same type in the world, reducing emissions of atmospheric pollutants. Through the use of noise reduction and temperature control technology, the team's remote point noise is less than 45 decibels, and the temperature rise within 500m around the circulating drainage point does not exceed 2 ℃. A protective layer is needed under the concrete to reduce the impact on groundwater. The ultra-supercritical clean coal-fired technology adopted by the project meets the EU's most stringent industrial carbon emission standards, which is conducive to local green environmental protection and energy conservation and emission reduction.

In addition, the project adopts high environmental protection requirements and pays attention to the protection of local ecosystems during the construction process. The construction area of the Dubai Haxiang project terminal is located in the local nature reserve. Harbin International actively fulfills its local social responsibility and environmental protection obligations. They transplanted and cultivated 28,000 corals in the construction area and transferred all fish in the lagoon to the sea. During the annual turtle breeding season, sea turtles are monitored and protected in accordance with a plan developed jointly with the Dubai Environmental Protection Organization. The construction party has set up a water quality and air quality monitoring station to monitor the environmental parameters of the construction area in real time. ①

---

① Source: http://big5. xinhuanet. com/gate/big5/big5. news. cn/gate/big5/silkroad. news. cn/2018/0727/105672. shtml.

IV. Development of the Green Climate Fund (GCF) and Analysis of Financing Options

According to the principle of "common but differentiated responsibilities" between developed and developing countries as determined by the United Nations Framework Convention on Climate Change(UNFCCC), developed countries should bear the primary responsibility for reducing greenhouse gas emissions. It is also obliged to provide new and additional financial support to developing countries. The Green Climate Fund(GCF) came into being under this principle.

(I) Development of the Green Climate Fund(GCF)

1. The GCF was established to raise funds for developing countries to cope with climate change

The Green Climate Fund(GCF) was set up by the 194 countries who are parties to the UNFCCC as part of the Convention's financial mechanism to provide funding for climate change mitigation and adaptation in developing countries.

The proposal to establish the GCF was first proposed at the 2009 Copenhagen Climate Change Conference. The Copenhagen Accord stipulates that developed countries will fund the establishment of a green climate fund to support developing countries in their response to climate change. The agreement preliminarily stipulates that developed countries should invest USD 30 billion as a quick start fund from 2010 to 2012, and provide USD 100 billion in annual support by 2020. The GCF was subsequently established at the Cancun Climate Change Conference in 2010. At the 2015 Paris Climate Change Conference, the United Nations reaffirmed the developed

IV. Development of the Green Climate Fund (GCF) and Analysis of Financing Options

countries' commitment to GCF. It is also proposed that technical experts will be hiredto assess the climate fund situation, and urged developed countries to complete the annual investment commitment of USD 100 billion, but they have not assigned mandatory funding tasks. The development of GCF is shown in the following figure:

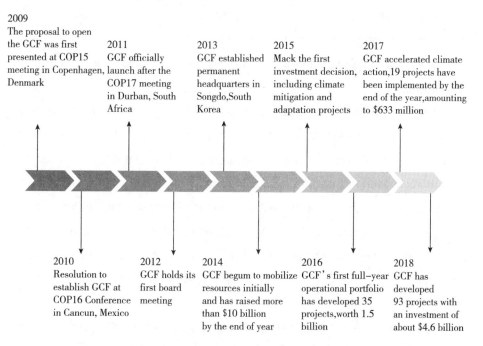

Figure 4 – 1   Green Climate Fund (GCF) Development Chart

2. The GCF has the Board, the Secretariat and the Trustee

The GCF has functional departments such as the Board, the Secretariat, and the Trustee. The main functions of GCF Board are to supervise the operation of the fund, approve business models, utilize models and funding structures, approve specific operational policies and guidelines, appoint the Executive Director of the Secretariat, select and appoint trustees, and so on.

The GCF Secretariat is appointed by the Board and is responsible to the Board. It is responsible for the day-to-day operations of the Fund and provides financial, administrative and legal support. In 2012, Incheon, South

Korea was selected as the seat of the Secretariat. South Korea pledged to inject USD 1 million annually intoGCF by 2019 and to provide USD 40 million to GCF in the form of a trust fund in 2014-2017.

The trustee must manage its financial assets in accordance with international standards and prepare financial statements and other relevant reports. Currently, the World Bank is a temporary trustee of the GCF and will be audited three years after the operation.

3. GCF's funding models are mainly grants, concessional loans, equity and guarantees

At present, GCF has raised a total of USD 10.3 billion from 43 countries. Its project investment has been carried out through the Accredited Entities(AE), and 75 entities have been accredited. These entities include international financial institutions and international organizations such as the World Bank, the United Nations Development Program, the United Nations Environment Program, and the Asian Development Bank. It also includes private sector institutions such as HSBC. Two of the entities are Chinese institutions, namely the China Clean Development Mechanism Fund Management Center and Foreign Economic Cooperation Office.

After the establishment of GCF, it has always attached importance to investment in climate adaptation projects and is committed to balancing the investment proportion of mitigation and adaptation projects. As of December 2018, 93 projects have been developed involving an investment of USD 4.6 billion and a cumulative reduction of 1.4 billion tons of CO_2 emissions. 39% are climate change mitigation projects, 25% are adaptation projects, and 36% are both involved. Geographically, the project is spread across major continents, with 40 projects in the Asia Pacificregion, 36 projects in Africa, 20 projects in Latin America and the Caribbean, and 6 projects in Eastern Europe. However, GCF has not yet invested in China.

IV. Development of the Green Climate Fund (GCF) and Analysis of Financing Options

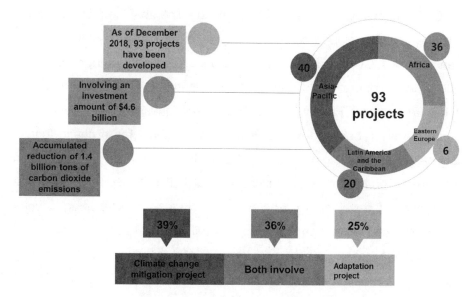

Figure 4–2 GCF Investment Project Status

GCF has four types of financing modes including grants, concessional loans, equity and guarantees for these projects. The investment amount was USD 2.1 billion, USD 1.9 billion, USD 420 million and USD 80 million respectively, accounting for 47%, 42%, 9% and 2% of the total investment.

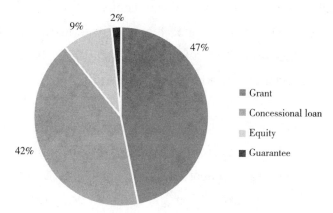

Figure 4–3 Proportion of each financing mode of GCF

4. Main features of GCF

(1) Focusing on equality principle in the operation of GCF

GCF focuses on the needs of areas most vulnerable to climate change, including least developed countries, small island developing States and African countries. GCF focuses on thestate parties' equality and geographical equality in its operations. Other funds under the UNFCCC, such as the Global Environment Facility (GEF), have been questioned by developing countries, mainly because of the operational mechanisms of these funds, marginal island countries and LDCs have almost no voice in investing in funds. In the course of its operations, the GCF fully took into account the participation of developing countries and included 25 developing country members in the preparation of the Transitional Committee, including 2 members from the least developed countries and small island developing States. There are 15 developed country members who give developing countries sufficient right to speak. After the formal establishment, the board of directors consists of 12 members from each of the developing and developed country Parties. The two co-chairs are elected internally by the board of directors, from developed countries to developing countries, fully embodying the principle of equality.

(2) Financing from public and private sectors

Compared to other climate funds that only raise funds from the public sector, GCF has set up a Private Sector Facility (PSF). The private sector can play an important role in addressing climate change, but private funds are more focus on profit rather than climate change, it is difficult to ensure that they are responsible for climate investments. GCF takes a cautious approach to absorbing private capital, fully assessing its impact on the environment, society and development, and ensuring its positive impact on mitigation and adaptation to climate change. Of the 93 projects GCF has invested, 40% are from the private sector. The PSF not only raises more funds for the GCF, but also drives the private sector in developing countries to participate in climate change mitigation and adaptation activities.

(3) The proportion of investment in developed countries is unresolved,

IV. Development of the Green Climate Fund (GCF) and Analysis of Financing Options

and GCF is exploring other financing channels

Although developed countries have pledged to finance GCF, their obligations are not mandatory, and there is a lack of specific and detailed funding requirements. The willingness of developed countries to invest is low, and they delayed the capital injection. According to the data from the official website of the Quick Start Fund, the developed countries promised to invest USD 30 billion in rapid start-up funds from 2010 to 2012, however, only USD 3.6 billion is completed.

Successive UN climate change conferences emphasized the importance ofestablishing GCF, the idea was also unanimously endorsed by the parties. However, there is still no unified statement and caliber for the proportion of capital contributions between developed countries. At the COP24 Polish Katowice Climate Change Conference, which was finished in December 2018, all Parties adopted the implementation rules of the Paris Agreement, provided a new climate funding method after 2025. Developed countries promised to provide USD 100 billion annually for developing countries in climate funding, the German and Norwegian governments pledged to double the investment in GCF, but the parties still did not reach the specific capital contribution plan for developed countries.

Therefore, in addition to the financial aid commitments of developed countries, GCF is currently actively mobilizing capable developing countries to inject capital, and promote cooperation between the public and private sectors, absorb social capital, and carry out project development.

(II) Exploring the Schemes for Green Climate Fund Financing

As a key issue in recent international climate summits, the Green Climate Fund (GCF) is confronted with the problem of insufficient financing. The GCF is a fundraising of international funds, which is sourced from developed countries and flows to developing countries. For international fund raising, there is a certain commonality in financing sharing. That is, due to the lack of an authority that transcends the sovereignty of each country, financing relies entirely on negotiation among countries, and the final plan is generally the result of compromise between countries. Therefore, it is necessary to consolidate and summarize the existing international fund raising model and explore its reference to GCF financing sharing.

The following contents will discuss how to share GCF financing tasks between developed countries. From the perspective of historical responsibility and ability to pay, we draw on the experience of existing international funds such as the United Nations Assessment(UN), the United Nations Official Development Assistance Program(ODA) and the Global Environment Facility (GEF). We discussed the allocation of GCF financing responsibilities under different financing mechanisms. Considering that different countries have different program preferences, voting theory is used to measure different plans and to obtain a PSC financing responsibility sharing system that considers multiple evaluation ideas. The paper will also analyze emerging economies' participation in climate finance scenarios and assess the impact of US exits on the Paris Agreement on climate finance development.

1. Single-principle schemes for financing the GCF

(1) Financing the GCF based on historical emission responsibility(HR)

Allocating finance responsibilities proportional to historical emissions is based on the idea that those countries with the highest contribution to the historical emission levels and the resulting climatic change should bear most of the climate finance responsibility, which is also defined as the HR approach. The cumulative CO_2 emissions from 1850 to 2015 in each region are selected as the basis for calculation. It is found that if GCF financing is based on the HR approach, the United States needs to contribute 44.30%, the EU needs to contribute 25.57%, the UK needs to contribute 8.36%, and Japan needs to contribute 6.82%. Under the HR approach, the US is under a lot of financial pressure and needs to bear nearly half of GCF funding obligations.

IV. Development of the Green Climate Fund (GCF) and Analysis of Financing Options

(2) Financing the GCF based on the ability-to-pay (AP) principle

The AP approach assumes that the GCF financing share of each country is linked to its economic capacity, and GDP is selected as a representation of economic capability. In order to eliminate fluctuations between years, the 2010-2015 average GDP is used as a basis for calculation. It turns out that, similar to HR, if the GCF fundraising task is based on the AP approach, the US, which sponsors 35.38% of the funds, is the largest contributor to the GCF, followed by the EU with a share of 30.38%. The share of donations from Japan and the United Kingdom is 11.67% and 6.37%, respectively. Other developed countries also need to undertake certain fundraising tasks, but the share is relatively small.

(3) Financing the GCF based on United Nations (UN) membership dues

The UN approach assumes that the share of contributions from Member States is proportional to the level of United Nations contributions, based on the 2013-2015 United Nations average annual contributions. As a result, if the GCF is financed based on the UN membership dues approach, the top 4 providers are the EU (36.61%), the USA (28.23%), Japan (13.90%), and the UK (6.65%). Unlike the results under HR and AP, the EU will be largest contribution to the GCF under UN.

(4) Financing the GCF based on Official Development Assistance (ODA)

ODA is an international aid program between developed and developing countries. The developed countries are the donor parties, whereas the developing countries are the recipients. This kind of aid flow is consistent with the setting of the GCF. We assume that the financing burden is allocated to each country, proportional to the national contribution to ODA. Based on the 5-year average of aid from 2010 to 2014, it is resulted that the EU contributes most with a share of 39.82%. The United States, the United Kingdom, and Japan contributes 22.81%, 11.40%, and 7.78%, respectively.

(5) Financing the GCF based on Global Environment Facility (GEF)

The GEF approach assumes that the GCF financing level of each funding country is proportional to its GEF funding level, that is, the countries that invest more in the GEF will be willing to undertake more green climate fund financing obligations. Under this approach, the EU is the largest contributor to the GCF, sponsoring nearly 43.47% of the funds. The US and Japan sponsor share is 14.83% and 16.49%, respectively. Compared with the above four approaches, the EU has the most financial pressure under GEF approach.

2. Financing the GCF based on the Preference Score Compromises (PSC) approach

(1) The design of PSC approach

All of the above five options can provide valuable lessons for GCF financing, but donor countries need to negotiate which one to adopt. In actual negotiations, different countries have different decision-making preferences for national interests. In order to reduce the differences between the funding countries in the GCF financing negotiations, it is necessary to seek a balance between different programs and to maximize the interests of different countries. Although all of the five proposed approaches (namely, HR, AP, UN, ODA, and GEF approaches) could present alternatives to financing the GCF, negotiations among the parties need to determine which alternatives to adopt. In fact, driven by national interest, each country tends to choose the scheme that provides the greatest benefit. Therefore, different donor countries may vary in their preferences. To solve conflicts between parties, balance among the five principles is necessary. These principles embody the interests of different countries as much as possible. Such balance may be achieved using the PSC approach.

Table 4-1 introduces the PSC method by using the alternatives (HR, GEF, and ODA approaches) as an example. The three alternatives, namely, HR, GEF, and ODA methods, are weighted by the PSC method to obtain a

IV. Development of the Green Climate Fund (GCF) and Analysis of Financing Options

composite approach for financing the GCF. Under the PSC approach, if a country prefers one alternative to the others, that alternative is assigned a score according to the population of that country. It can be seen that the preference weights of the three alternatives in developed countries are 47.09%, 47.70% and 5.21%, respectively. Under PSC, the financing share of the US is 30.34% and the EU is the largest contributor to the GCF, sponsoring 35.05% of the funds. Japan and the UK provide 11.85% and 8.51% of the financing, respectively. The remainder is divided among the other developed parties. For example, Canada and Australia should finance 4.36% and 1.93% of the funds.

Table 4-1 An example of the PSC approach (alternatives: HR, GEF, and ODA approaches)

	Alternatives(%)			Voting process(in millions)			Results(%)
	HR Approach	GEF Approach	ODA Approach	Vote for HR Approach	Vote for GEF Approach	Vote for ODA Approach	Burden-sharing
USA	44.33	28.23	16.76	0.00	0.00	315.47	30.34
Japan	6.82	13.90	16.59	127.47	0.00	0.00	11.85
Canada	3.46	3.83	5.30	34.94	0.00	0.00	4.36
Australia	1.86	2.10	1.99	23.00	0.00	0.00	1.93
NZW	0.20	0.32	0.21	4.46	0.00	0.00	0.21
Switzerland	0.32	1.34	3.20	8.06	0.00	0.00	1.75
Norway	0.29	1.09	1.73	5.05	0.00	0.00	1.02
UK	8.37	6.65	8.86	0.00	63.93	0.00	8.51
EU	25.59	36.61	44.17	374.28	0.00	0.00	35.03
SKA	1.71	2.56	0.24	0.00	0.00	50.08	1.05
Mexico	2.04	2.36	0.35	0.00	0.00	119.63	1.25
ODV	5.02	1.01	0.60	0.00	0.00	99.60	2.70
Total	–	–	–	577.26	63.93	584.78	100.00

Note1: ODV is "other developed countries", including Chile, Panama, Iceland, Colombia, Liechtenstein, Peru and Monaco.

Note2: Assume the alternatives are HR, GEF and ODA.

(2) Exploring the schemes for financing the GCF under the PSC approach

The PSC approach is essentially a multi-index weighting method. If HR, AP, UN, GEF and ODA are alternatives, and different combinations of different schemes are allowed, by permutation and combination, the set {HR, AP, UN, GEF, ODA} contains 31 nonempty subsets, as shown in Figure 4 – 4. It can be found that although the differences between the different schemes are obvious, the United States, the European Union, Japan, and the United Kingdom have always been the most important contributors, and they have to contribute more than 80% of the funds. This indicates that the GCF is very dependent on the above four regions.

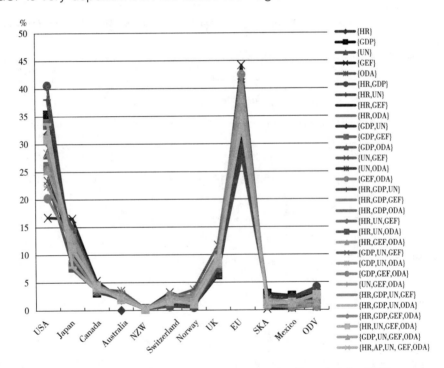

Figure 4 – 4 Results of potential schemes for financing the GCF

3. Evaluation of the effectiveness of the GCF financing sharing schemes

In order to evaluate the effectiveness of the 31 potential schemes, GCF's first fund raising period (2015-2018) was selected as the Benchmark and using the methods of Euclidean distance and Correlation coefficient, the

IV. Development of the Green Climate Fund(GCF) and Analysis of Financing Options

results are shown in Figure 4 -5. It can be seen that the effectiveness ranking of the potential schemes under the two methods is exactly the same, indicating that the results have strong robustness. In the Euclidean distance scenario, the closest scheme to the Benchmark is {HR, GEF, ODA}, and the difference between them is 0.06. In the Correlation coefficient scenario, the scheme most similar to the Benchmark is also {HR, GEF, ODA}, and the correlation coefficient is as high as 0.99, which indicates a highly positive correlation between the two. In addition, it is found that the most dissimilar or least close scenario is the HR scheme, which indicates that GCF financing allocation should not only consider the historical emissions responsibility of each country, but also balance the economic capabilities of all parties.

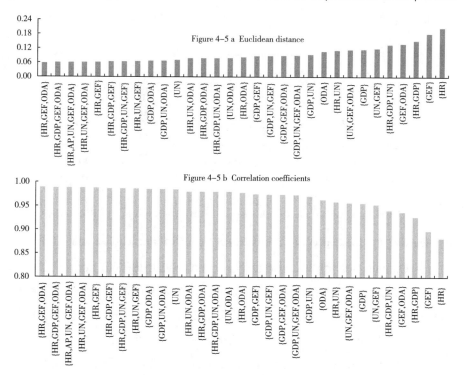

Figure 4 – 5 Effectiveness ranking of 31 design schemes(the upper graph is the Euclidean distance metric, and the lower graph is the correlation coefficient metric)

4. Sensitivity analysis

(1) The impact of the US withdrawing climate finance

This section mainly discusses the impact on the other donor countries after the US withdrawing climate finance, and analyzes the most effective {HR, GEF, ODA} scheme as an example. As shown in Figure 4-6, the withdrawal of US climate financing will increase the burden of donations from other countries, especially for the EU. Specifically, the EU needs to undertake 48.83% of the fundraising task, Japan needs to undertake 15.53% of the fundraising task, and the UK needs to undertake 12.97% of the fundraising task. Compared with the scenario that the United States do not withdraw (Table 4-1), the EU's share of investment increased by 13.80%, the United Kingdom increased by 4.46%, and Japan increased by 3.68%.

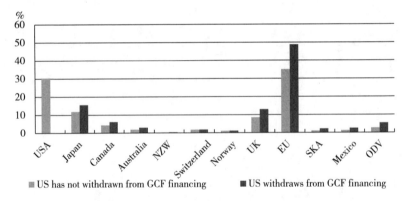

Figure 4-6 The impact of US withdrawal from GCF financing on other donors

(2) Climate finance members are broadened to BRICS countries

This section discusses the impact of GCF donors' expansion to the BRICS countries, as well as {HR, GEF, ODA}. As shown in Figure 4-7, even if GCF financing members are broadened to BRICS countries, developed countries are still the main donors of the Green Climate Fund, and they need to undertake 94.43% of the funding tasks. The BRICS countries need to assume 5.57% of the financing obligations. At this time, China needs to contribute 2.54%, Russia needs to contribute 1.36%, Brazil needs

IV. Development of the Green Climate Fund (GCF) and Analysis of Financing Options

to contribute 0.51%, and India and South Africa need to contribute 0.85% and 0.31% respectively. This shows that the {HR,GEF,ODA} program will not bring obvious financial burden to the BRICS countries, further showing the rationality of the design of the program.

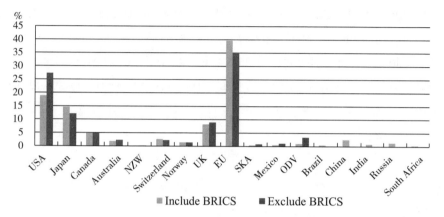

Figure 4-7 **The impact of broadening GCF financing to BRICS countries**

5. Conclusions

(1) The responsibility sharing of different financing mechanisms varies greatly. For example, the United States has the smallest share of funds under the GEF program, and Japan and the EU share the smallest share under the HR principle.

(2) Using PSC method to combine HR, UN and GEF can get the most effective one. The apportion effect is closer to reality. At this time, the EU needs to invest 35.03%, the US needs to invest 30.34%, and Japan and the UK need to invest 11.85% and 8.51% respectively.

(3) If climate finance members are broadened to the BRICS countries, the region will have to undertake a total of 5.57% of funding tasks, indicating that GCF donations are more dependent on developed countries.

(4) The withdrawal of the United States from the Paris Agreement will significantly increase the financing burden of other countries, and the EU's share of contributions will increase by 14%.

Therefore, sharing of GCF financing responsibility needs to consider not only historical emission responsibilities, but also the economic capabilities of each donor country. The trade-off between the two needs to take into account the preference of different countries, which helps to improve the acceptability of the program. This research proposes to use the PSC method to compound HR, GEF and ODA, and construct a distribution plan for GCF financing obligations. This is not only close to the regional contribution of GCF's first capital increase, but also does not impose an excessive financial burden on emerging economies. It is a balance between fairness and efficiency.

V. The Role of Multilateral Development Banks in Climate Finance

Following Monterrey, Mexico, 2002, and Doha, Qatar, 2008, the third United Nations International Conference on Financing for Development was held in Addis Ababa, Ethiopia in 2015. "We encourage the multilateral development finance institutions to establish a process to examine their own role, scale and functioning to enable them to adapt and be fully responsive to the sustainable development agenda." [1]

Given their mandate, size, and influence, MDBs play a critical role in the reaching the SDGs. MDBs bridge the gap from public to private by directing for-profit driven financial markets towards policy objectives[2]. Through their characteristics, they can incentivize and demonstrate the feasibility of certain methods, sectors, or geographies, that other investors may not otherwise be inclined towards. Working towards policy objectives, MDBs have increased their involvement in climate finance both by scaling up climate investments and by integrating environmental issues into their general financing requirements as a cross-cutting issue. To increase these efforts the G20 Leaders Statement in Hamburg calls for MDBs to enhance climate activities[3]. Furthermore, MDBs issued a joint statement at the One Planet summit in Paris in December 2017, promising to align their financial flows with

---

[1] Addis Ababa Action Agenda. Addis Ababa Action Agenda (AAAA) of the Third International Conference on Financing for Development. Addis Ababa, Ethiopia: United Nations [R]. 2017.

[2] MDBs. From Billions to Trillions: MDB Contributions to Financing for Development [R]. 2017.

[3] G20. G20 Leaders Statement 'Shaping an interconnected world'. Hamburg, Germany: G20 [R]. 2017.

the Paris Agreement[①].

(Ⅰ) MDBs has advantages in promoting sustainable development

Following Monterrey, Mexico, 2002, and Doha, Qatar, 2008, the third United Nations International Conference on Financing for Development was held in Addis Ababa, Ethiopia in 2015. This event brought together a wide array of stakeholders to discuss how to finance the SDGs, which were being settled around the same time in a parallel process. Participants at the event included more than 50 heads of state and 200 ministers, UN and other intergovernmental institutions such as the WTO, prominent businesses, civil society organizations and other stakeholders. The outcome of the document was the Addis Ababa Action Agenda (AAAA).

In contributing to sustainable development, the AAAA highlights five primary characteristics of MDBs[②], namely 1) long-term & stable, 2) counter-cyclical, 3) concessional terms, 4) know-how & technical assistance, 5) private capital mobilization.

The following table summarizes the characteristics, advantages and typical examples of multilateral development banks:

Table 5-1 Summary of Characteristics of MDBs and their Advantages

MDB Characteristic	Comparative advantage	Representative examples
Long-term and stable	MDBs make long term commitments to projects, providing stable investment at both project, sector, and geographic level	Average MDB loan maturity of 20-30 years The WB and ADB's long term commitment to improving urban air quality in Asia through the Clean Air Initiative

---

① MDBs. Joint Statement by the Multilateral Development Banks at Paris, COP21. Paris, France: MDBs[R]. 2017.

② Addis Ababa Action Agenda. Addis Ababa Action Agenda (AAAA) of the Third International Conference on Financing for Development. Addis Ababa, Ethiopia: United Nations[R]. 2015.

V. The Role of Multilateral Development Banks in Climate Finance

continued

MDB Characteristic	Comparative advantage	Representative examples
Counter-cyclical	Willingness to invest in strategic areas under circumstances where other investors are pulling back engagements	MDBs statistically scaling up of climate financing in the wake of the global financial crisis, offsetting national budget cuts in many countries[1]
Concessional terms	Ability to provide concessional terms in a variety forms including blended financing towards strategic areas	The ADB's climate finance catalyzing facility or the IaDBs green credit lines, aiming to increase the bankability of green projects
Know-how and technical assistance	Large scale and long history of MDBs provide greater overarching expertise than other investors for both the specific project, and the country or region as a whole	The EIB and WB played a catalytic role in promoting green bonds, through their legitimacy and ambitious engagements. The new IMF-WBG partnership on tax diagnostics
Private capital mobilization (financial policy supportand specialized mechanisms)	Through financial policy support and financial mechanisms MDBs can reduce perceived risks at both institutional and project level	Whereas MDBs catalyze USD 2-5 of private capital per USD 1 invested[2], the average for public north-south climate financing is USD 0.34[3]

(Ⅱ) Challenges within climate finance

The unmet need for climate financing is fundamentally a sign of a lower return on investment from a combination of actual and/or perceived lower revenues and higher costs – as based on numerous factors inside such calculations. On the revenue side, low-carbon investments often have a sim-

---

[1] OECD. "CRS:Aid activities", OECD International Development Statistics(database). Paris, France: OECD[R]. 2016.

[2] MDBs. From Billions to Trillions: MDB Contributions to Financing for Development[R]. 2015.

[3] OECD & CPI. Climate Finance in 2013-2014 and the USD 100 billion goal. Paris, France: OECD[R]. 2015.

ilar revenue as non-climate projects, or it may even receive targeted subsidies such as a feed-in tariff for electricity generation. However, it is particularly on the cost side that low-carbon projects are perceived as less attractive to investors. These costs often stem from higher technological risk within less mature technologies, political risks from less stable political support or the lack thereof, higher implementation costs from using less proven implementation methods, mismatch of project and lending timelines, and higher transaction costs in financial markets.

An overview of the challenges facing climate finance development is given in the table below:

Table 5-2 Summary of Challenges in Climate Finance Development

Level of Challenge	Key Aspects	Examples of Current Overarching Efforts
Institutional Framework	A. Political, economic, & environmental instability B. Policy reversals & regulatory uncertainties C. Distorting subsidies and feed-in tariffs D. Uneven playing field to SOEs E. Regulatory barriers to entry	Streamlining of policy implementation from strategic to regulatory and local level including via international fora and mechanisms
Project Financiers	A. High project development costs B. High transaction costs C. Competition for low-carbon projects between providers of specialized funds D. Overemphasis on short-term returns E. Portfolio restrictions	Compensation mechanisms for increased costs of launching sustainable projects, improving ESG awareness of investors, and incorporating climate finance requirements

Ⅴ. The Role of Multilateral Development Banks in Climate Finance

continued

Level of Challenge	Key Aspects	Examples of Current Overarching Efforts
Project Owners	A. Limited awareness of climate finance mechanisms B. Inexperience in leveraging non-traditional finance C. Limited capacity for structuring projects as low-carbon D. Lack of publication of transparent and comprehensive project pipelines E. Lack of viable funding and business models	Improving guidance and training available as provided by public or private organizations, and developing platforms for knowledge exchange and financing mechanisms
Financial Markets	A. Lack of low-carbon asset classes B. Shortage of specialized funds and tools C. Mismatch in risk profiles D. Non-monetized positive environmental externalities E. Lack of data F. Low ability to accurately assess low-carbon project risks.	Improving information availability and quality through platforms, standards, and third-party assessments, developing innovative financial tools addressing, and establishing international dialogue on climate finance harmonization of processes and standards

(Ⅲ) MDBs are playing important roles in climate finance

In 2016 all MDBs together provided over USD 27 billion for climate financing, of which 77% were labelled as mitigation and 23% as adaptation. In order to achieve the 2020 UNFCCC 2020 commitment target, the multilateral development banks will provide strong climate financing support between 2013 and 2015, with support quotas accounting for more than one third of the total support for developed countries to developing countries. Although the cumulative support amount of multilateral development banks is

185

relatively large, it is also important to consider the proportion of climate finance in the total financing of multilateral development banks, and also need to consider the climate share of multilateral development banks in non-climate portfolios.

The figure below shows a number of the MDBs climate financing proportion compared to 2020 targets. According to the commitment, the multilateral development bank will provide 40% of the total developed countries' climate finance support to developing countries. This shows that as most MDBs need to make drastic progress within a short period to achieve their targets, they have to carry out rapid change in financing direction compared to business-as-usual. In addition to active targets for climate finance, MDBs are also using negative lists for excluding projects. For example, the WB President Jim Kim announced at the One Planet Summit in Paris 2017 said that the WB will no longer finance upstream oil and gas, while the bank quit coal fired plants in 2010.

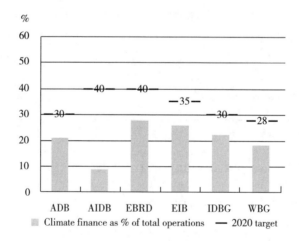

Data Source: MDB Climate Finance. The Good, the Bad and the Urgent. Washington DC, USA: WRI[R]. 2017.

Figure 5 – 1　MDBs 2016 Climate Finance Proportion and 2020 Targets

It is further important to consider the overall climate financing of MDBs portfolios. As mentioned above, the World Resources Institute has analyzed

V. The Role of Multilateral Development Banks in Climate Finance

the climate financing by the World Bank(WB), International Finance Corporation(IFC), and Asian Development Bank(ADB) (accounting for 1/3 of MDB financing), compared with a 2℃ scenario. They conclude that 17% of financing is aligned with a 2℃ pathway, 57% is conditional, 22% is controversial, and 3% are misaligned[①]. Despite the study's narrow coverage, it provides an indication that MDBs need to change business-as-usual to be aligned with climate policy objectives. This is further highlighted by the OECD's calculations, as shown in the figure below. From this figure, it is clear that while climate change has been incorporated into the energy sector of infrastructure, it remains a much smaller proportion in other sectors. In total, a third of all MDB infrastructure financing in 2013-2015 targeted climate change mitigation or adaptation.

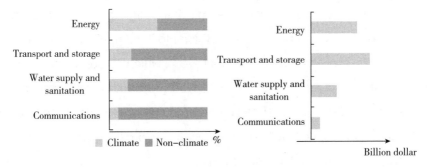

Data Source: OECD. Investing in Climate. Investing in Growth[R]. 2017.

Figure 5 – 2 Climate Change as Proportion of MDB Infrastructure Financing by Sector
(Left: Proportion Right: Quantity)

(Ⅳ) MDBs should continue to promote Climate Finance in various ways

There is a clear trend of increasing importance of climate finance amongst MDBs. This is shown clearly in strategic documents such as long-term plans, in promises to reach a minimum proportion of the total invest-

---

① WRI. Financing the Energy Transition: Are World Bank, IFC, And ADB Energy Supply Investments Supporting A Low-Carbon Future?. Washington DC, USA: WRI[R]. 2017.

ment stock in climate related fields, as well as in the rapid increase in MDB's green bond issuances. While early MDA base licensing documents focused on development, growth and transformation, they allowed the use of current strategy documents to increase the volume of greenbusiness. The newer established MDBs often have green as a central pillar embedded in their original mandate.

The multilateral development banks can consider the following measures to further their role in the field of climate finance:

1. Prioritization of Private Capital Mobilization

MDBs can play a critical role in mobilizing the private capital required to finance environmental sustainability. It is acknowledged that the private sector will provide a critical proportion of financing for sustainable development and that MDBs historically have one of the highest capital leverage abilities. According to the MDB's From Billions to Trillions Report, mobilization rates of MDBs are often between USD 2-5[①], which is substantially higher than estimates of other sources such as North to South climate financing[②]. While some MDBs do so already, more multilateral development banks should be involved and use it as the basis for all businesses.

2. Implement Internal Carbon Pricing

Since external carbon pricing is not applied in all countries, MDBs can use internal carbon pricing to increasingly internalize externalities and mitigate physical and transition risks in project financing. Depending on the method of internal carbon pricing this can directly impact project bankability or at the minimum provide a basis for discussing the carbon footprint of each project. To be compatible with the Paris Agreement, the Carbon Pri-

---

① MDBs(2015). From Billions to Trillions: MDB Contributions to Financing for Development.
② OECD & CPI (2015). Climate Finance in 2013-14 and the USD 100 billion goal. Paris, France: OECD.

cing Leadership Coalition suggests a USD 40-USD 80 range by 2020[①]. While internal carbon pricing can provide a direct cost incentive in project evaluation, it can also be used in tandem with improved environmental risk assessment methodologies to comprehensively capture environmental factors on both the cost and risk side.

3. Focus on Targeted Rather Than Broad Concessional Support

While MDBs' concessional financing can incentivize financing towards certain policy objectives, such support may have adverse market distorting effects. MDBs should carefully target their concessional support in climate finance rather than provide broad subsidies, along the guidelines of the Development Finance Institutions' recommendations. Many cases in climate finance show that project investors need short-term preferential interest or preferential terms and technical support from multilateral development banks compared to long-term preferential financing.

4. Enhance Environmental Risk Assessment

Despite MDBs have sufficient finance experience, the rapidly changing environmental circumstances presents challenges to current methodologies of risk assessments. In addition to traditional project internal financial properties and external risk factors, MDBs have to use new methods for including physical climate risks and transitional environment-related risks as highlighted by the 2017 G20 climate finance Study Group[②] and the Financial Stability Board's Task Force on Climate Related Financial Disclosure[③]. Such methodologies should adequately include asset-level data, impact measurement, potential scenarios, management implications, and other

---

① Carbon Pricing Leadership Coalition. Report of the High-Level Commission on Carbon Prices[R]. 2017.
② G20 Hamburg. G20 climate finance Synthesis Report 2017. Hamburg, Germany: G20[R]. 2017.
③ FSB-TCFD. Recommendations of the Task Force on Climate-related Financial Disclosures. Basel, Switzerland: FB[R]. 2017.

case-specific variables.

5. Expand MDB Cooperation for Economies of Scale

MDBs can increase the efficiency of their climate financing by merging together a number of financing solutions across MDBs. While MDB competition on policy advice, pricing, and financing modalities can be healthy it can also lead to a suboptimal outcome for the development finance system[1]. If there are many overlaps in the different financing solutions of the multilateral development banks, then economies of scale can be used to improve efficiency and ultimately achieve the partnership principle under the Busan Effective Development Partnership[2].

6. Extend Reapplication of Non-Climate Financial Solutions as Climate

As MDBs are increasingly emphasizing financing for climate areas, existing successful financing solutions targeted at other priorities can in some cases be duplicated or modified to include climate finance. Such an update on policy and practice towards the post-2015 development agenda is directly encouraged by theAddis Ababa Action Agenda[3]. Each MDB can analyze their respective financing solutions, to determine low-hanging fruits for scaling up climate financing towards their goals.

---

[1] Brookings. The New Global Agenda and the Future of the Multilateral Development Banking System. Washington DC, USD: Brooking Institution[R]. 2018.

[2] OECD. The Busan Partnership for Effective Development Co-operation[R/OL]. 2011. Retrieved from: http://www.oecd.org/development/effectiveness/busanpartnership.htm.

[3] Addis Ababa Action Agenda. Addis Ababa Action Agenda(AAAA) of the Third International Conference on Financing for Development. Addis Ababa, Ethiopia: United Nations[R]. 2015.

VI. Policy Recommendations

(Ⅰ) Continue to play the role of 'leader' under the new global climate governance

After the Paris Agreement, the global climate governance pattern has changed. From the mandatory allocation of "top-down" greenhouse gas emission reductions advocated by the European Union to the "bottom-up" Intended Nationally Determined Contributions (INDC) mode advocated by China and the United States. INDC reflects the commitments and specific actions that each country is willing to undertake. It is a relatively loose and flexible greenhouse gas emission reduction model. It is more suitable for the current global climate change situation and is conducive to the smooth progress of climate negotiations.

At present, the EU's climate governance leadership has gradually weakened, the US Trump administration has announced its withdrawal from the Paris Agreement, and China has changed from a "follower" to a "leader" in the climate governance system. During the Copenhagen conference, China became an important participant in international climate negotiations and voiced on behalf of developing countries. Afterwards, the rapid rise of the "basic four countries" represented by China played a key role in the trend of climate negotiations and the transformation of governance models, and adhered to the basic principle of "common but differentiated responsibilities" and implemented them into In the Paris Agreement. During the "Paris Agreement" period, China not only proposed actual emission reduction targets in its own contribution documents, but also reflected the responsibility of major powers, and coordinated other major countries to jointly promote the signing of the Paris Agreement. In September 2016, China took advantage of the home court advantage of the G20 summit to facilitate the formal entry of China and the United States into the Paris Agreement, which greatly promoted the entry into force and implementation of the agreement. In December 2018, the Katowice Climate Change Conference completed the negotiation of most of the implementation rules of the Paris Agreement and achieved strong results. China has made important contributions to the success of the conference and has become a key driving force for international climate negotiations.

In the context of the implementation of most of the contents of the Paris Agreement, and in the context of the shift in climate policy in the United States due to government changes, China has adhered to the responsibility of major powers in climate governance. And it is a wise strategic choice for China to follow the trend of development and form stronger international leadership. It is conducive to leading the development of international rules and improving the right to speak on international affairs, improving the ability to deal with international affairs and learning from internal experience to solve internal affairs, which is conducive to promoting cooperation with other countries and further increasing China's influence on developing countries. It meets the needs of domestic industrial structure upgrading and environmental governance.

In order to play a leading role in tackling climate change, China should adhere to the following principles:

First, China should insist flexible and pragmatic attitude, and create multi-win situation for all parties involved. China should also adopt a positive and confident climate foreign policy, cooperate with developed countries and developing countries in an open manner, at the same time be more flexible and pragmatic, and not engage in confrontation between developing and developed countries on climate issues. Instead, China should seek

VI. Policy Recommendations

more oppotunities from a global perspective and play a leading role in tackling and leading global climate change issues on the international stage, maximizing the interests of China and other countries, and seeking multi-win.

Second, China should take advantage of the trend and establish a responsible country image. China and the United States are the largest carbon emitters and have taken on greater pressure to reduce emissions. At present, as a developed country, the United States has changed climate policy because of government changes, and has become a passive force in the global response to climate change. The United States is the target of public criticism, China needs to continue to adhere to its principles of climate foreign policy in recent years. Guided by the situation, it can achieve the effect of exerting stronger international leadership without additional practical input. This is an opportunity for China to lead climate change, actively participate in and influence international affairs, and establish its own image. It is worth mentioning that it is necessary to avoid blindly providing funds to reflect the responsibility of the major powers, and to adhere to the "common but differentiated responsibilities", on the one hand to maximize China's interests in the international game, and on the other hand to properly reflect China's responsibility.

Third, China should attach importance to strategic cooperation and bilateral cooperation. The "Basic Four Countries" cooperation and "South-South cooperation" are the main support for China's building leadership. Strengthening the cooperation of the "Basic Four Countries" is conducive to enhancing the sharing of climate information, coordinating the negotiation position, and deepening pragmatic cooperation in new energy emission reduction technologies. The "South-South cooperation" could be fully integrated with Belt and Road Initiative, which is conducive to exporting China's production capacity in new energy, it can also be combined with the investment layout of the AIIB to carry out low-carbon investment. Under the framework of multilateral cooperation, it is recommended to attach great importance to bilateral cooperation with the "basic four countries" and Belt and Road countries, and to carry out in-depth cooperation in green finance areas such as infrastructure investment and new energy investment that are conducive to emission reduction and environmental protection.

(Ⅱ) Actively use international climate funds to mitigate and adapt to climate change

1. Standing on the stand of developing countries to promote the realization of the "USD 100 billion" goal

The "USD 100 billion" climate fund should be "new, extra" funds provided by developed countries to developing countries under the UNFCCC framework, and should be based on public funds. China should unite with other developing countries, adhere to this principle, motivate the international consensus as soon as possible, and make the "USD 100 billion" goal a measured basis.

In addition, China needs to adhere to its own position as a developing country and adhere to the position of China's fund recipients. International climate funds should not be confused with domestic investment in autonomous emission reductions in developing countries. It is unwise to deprive China of the right to grant funds because of China's achievements in independent emission reduction. China's climate risks are still high, and the funding gap in the area of mitigation and adaptation is huge, and there is no doubt that international funding and technology support is needed.

2. Enhance own capacity building and strengthen the monitoring and statistics of recipient data

At present, the tracking reports on climate funds are from developed countries or the United Nations. As recipients of climate funds, there are no official reports from developing countries that track and monitor climate flows. Although developing countries seriously question the credibility of climate fund data published by developed countries, they cannot distinguish whether the acceptance data of climate funds are correct or not because they do not have their own statistics. China should participate in the construction of the International Climate Funds Verification, Reporting and Monitoring (MRV) system and align other developing countries to establish a climate fund MRV system from the perspective of receiving countries. This will not only help improve the international statistical system, but also provide a strong basis for developing countries in international climate negotiations. With its own statistics, China can better challenge and question the problematic data. In addition, the establishment and improvement of the climate fund MRV system can help China to clarify the funding gap, understand the gap between fund coordination and management, and help to formulate climate finance strategies and policies that are more in line with national conditions at the national level.

3. Pay attention to the demonstration effect of GCF and learn from the rich experience of international climate governance

At present, GCF has not invested in projects in China. China should actively seek financial support from GCF. Among the 75 implementing agencies of GCF, there are already two Chinese institutions that have the conditions to apply for financial support through domestic institutions. Although the funds available from GCF are relatively limited, its demonstration effect is very prominent, especially for the capacity building of backward areas, it can introduce the international rich knowledge and experience through project implementation, and strengthen the cooperation between international and domestic capital. In addition, through the GCF demonstration project, it can share and disseminate its advanced experience in climate management, provide excellent examples for other climate governance projects in China, and improve the climate change responding ability of the project location.

GCF commonly invests in the form of grants, concessional loans, equity and guarantees, of which grants and concessional loans accounts for more than 90% of the total investment. China could consider cooperating with GCF in these forms and explore new investment methods such as bonds to attract international funds to invest in China's climate finance sector. Given that GCF only funds for projects with good inherent income-generating projects, China should consider applying for potential other public or private funds as well as GCF funds to expand the level of financing preferences for GCF's application program.

4. Actively influence the formulation of operational rules and technical rules for the international climate finance mechanism represented by GCF

How to balance the allocation of funds is the biggest challenge GCF is currently facing. Balanced distribution is reflected in the balanced distribution of funds in the areas of adaptation and mitigation, as well as equitable distribution in developing countries.

China should actively influence the formulation of rules for the international climate finance mechanism represented by GCF, and avoid the developed countriesand international organizations representing their interests to formulate international rules in a one-sided manner. The relevant technical rules in climate funds are an important manifestation of the nature of climate funds in the operation of capital entities. On the basis of adhering to the independence of the climate finance such as GCF, the use effect and the continuous long-term funding, the technical standards become the rules that actually affect what kind of climate funds and their quantities can be obtained by developing countries. China should actively influence the formulation of rules, rationally voice on international occasions, and insist that such technical standards should reflect the universal will of developing countries and ensure the balanced distribution of specific funds.

(Ⅲ) Vigorously promoting domestic carbon market and developing carbon finance

1. Promote the healthy and stable development of the carbon spot market

The development of the carbon spot market is the cornerstone of carbon futures and other carbonderivatives innovations. Its healthy and stable development depends on the maturity of the following conditions:

The first is a perfect market top design. The legislative procedures for carbon market construction should be completed as soon as possible, relevant supporting systems should be improved, and scientific, unified and enforceable methodologies should be established. Besides, infrastructure construction should be done, data accuracy and information transparency should be improved to make full use of cap-and-trade mechanism, truly play the role of market pricing to achieve emission reduction targets.

The second is a stable and active trading policy. The policy should be stable and clear, such as the allowance issuance mechanism, CCER policy, etc. Then it is necessary to encourage transactions with a certain scale, in order to form a reasonable carbon price. In addition, diversified market participants should be introduced. Reasonable market size and sufficient liquidity is the basis for the carbon market to effectively play its pricing function. Individual investors should be cautiously introduced when the market is not mature, because they might blindly enter the market and increase market volatility.

The third is a flexible price control mechanism. It is suggested to establish a price stability control mechanism for the carbon market. We can establish a government reserve pool and a carbon market leveling fund to rationally set the carbon price control interval, and establish an open market operation mechanism based on the carbon allowance reserve and the leveling fund. It is also suggested to set price limit and maximum position limit to limit the maximum percentage of the day's rise and fall and the maximum position of the allowance holder.

2. Promote the construction of carbon futures market and access to financial supervision

The futures market is an effective complement to the spot market, it has the function of risk management, hedging, price discovery, providing liquidity and arbitrage opportunities. The carbon futures market can allocate carbon assets more effectively. Therefore, it is worthy to start research on the construction of carbon futures market, expand the hierarchical system of carbon market, realizes mutually beneficial interaction between carbon spot market and futures market, and can promote the good construction of carbon emissions trading system.

At the same time, financial supervision should be strengthened. Both macro andmicro levels should be prudently supervised and comprehensive preventive measures should be implemented to ensure the risk prevention and reduce the risk amplification. As the China Securities Regulatory Commission(CSRC) has extensive experience in market mechanism, price regulation, future derivatives, risk prevention and control, etc., it is recommended to introduce CSRC into the supervision and conduct joint supervision.

3. Strengthen stakeholders' carbon finance capacity building

Stakeholders mainly include governments, enterprises and financial institutions. The carbon market is an artificial regulation rather than a self-generated market. It is greatly influenced by policies. Therefore, as a policy maker, the government should be equipped with a certain carbon financial knowledge reserve, so that it can grasp the extent when formulating policies, not only ensuring the stable operation of the carbon market, avoiding excessive speculation, but also increasing the activity of the carbon market. As the main players in the carbon market, enterprises and financial institutions can only promote the development and promotion of carbon financial derivatives by improving their carbon finance capabilities, thus stimulating the healthy and healthy development of the entire market.

There is a big gap in the professional talents of carbon finance compared with market demand. It is necessary to design a comprehensive carbon finance course, introduce authoritative teachers, and regularly train stakeholders such as the government, enterprises, and financial institutions to familiarize themselves with the business processes and rules, and cultivate the theoretical knowledge and technical level of the incumbents. In addition, the propaganda should be strengthened to change the traditional ideas, enhance social awareness of energy conservation and emissions reduction and carbon finance as a whole.

4. Encourage pilot of carbon finance products innovation

In order to promote the hierarchical development of the carbon finance system, the government can take incentive measures to appropriately encourage the pilot of carbon financial product innovation on the basis of steady progress. Financial institutions should enrich their product types, encourage innovative forms, and explore carbon finance products that are suitable for the actual needs of enterprises and institutions, increase the operability, simplicity, and ability to flow of carbon finance derivatives. For high-risk carbon finance derivatives, OTC products, such as forward and swap products can be promoted in the near future to gradually promote the launch of on-market financial products, such as futures and options products, and pay attention to effective supervision and risk prevention.

(Ⅳ) Strengthening climate investment and financing along the Belt and Road

1. Continue to promote infrastructure construction that is conducive to climate change

Backward infrastructure construction and low levels of social and economic development are the common problems faced by BRI countries. Therefore, investments in infrastructure construction that is conducive to climate changes can not only lay a foundation for the economic development of these countries and solve their urgent needs, but also contribute to the protection of their ecological environment. Green and low-carbon construction can be achieved through attaching services and supports on protecting water, atmosphere, soil and biological diversity to infrastructure projects, as well as promoting clean transportation, clean energy and green building projects (e. g. railway, urban railway, urban and rural road).

2. Increase investment in clean energy

BRI countries possess enormous renewable energy potentials, including wind, solar, water and tidal energy, and therefore, leave huge rooms for development. For example, many countries in Central Asia and West Asia have long sunshine time, high light intensity and rich potential for solar energy development, as well as vast areas and abundant wind energy for wind turbine. Many countries in Southeast Asia and South Asia have water resources, such as waterfalls and rivers, for hydropower generation. By increasing investments in clean energy and giving full play to the energy advantages of BRI countries, the energy shortages of these countries can be largely alleviated while being environmentally friendly.

3. Promote investment and financing mode innovation

At present, China's investment projects in BRI countries are mainly on clean transportation infrastructure and clean energy, which normally have long construction and investment cycles. It is difficult to obtain investment returns in a short time when adopting the traditional equity financing. Therefore, it should be an effective measure to attract more investments by exploring new financing modes and maximizing the combination of long-term and short-term returns on the premise of controlling investment risks. For example, the innovative "equity + debt" financing mode of the Silk Road Fund, where share subscription and issuance of loans take place at the same time, balances the risks and returns of the investment as well as long-term and short-term returns.

4. Establish a Belt and Road Climate Investment and Financing Statistics System

So far, China has carried out many climate financing projects in BRI countries, which not only equip these countries with capacity to cope with climate changes but also help mitigate energy shortages and transportation inconvenience. To provide data support for international climate change negotiations and facilitate more targeted BRI climate finance, China should establish a statistical system recording these projects with concrete statistics and special management. By doing so, the projects with significant climate effects can be publicized, indirectly setting a positive example for other countries.

5. Strengthen the construction of climate cooperation platform

China should promote international climate cooperation through existing platforms (e. g. the SCO, the Forum on China-Africa Cooperation and the Lancang-Mekong Cooperation Mechanism) and provide supportive services for relevant projects. By giving full play to the existing bilateral and multilateral international cooperation mechanisms on environmental protection, a climate cooperation network can be built in certain regions to facilitate related countries in climate governance. In addition, for the proposed international climate governance system, China should take the lead in exploring comparative advantages of all parties by building a cooperation platform involving governments, think tanks, enterprises, social organizations and the public.